This book dedicated to the

....

This book dedicated to Mr.Arun kumar Kar (Manager in I.O.C.L) and Dr. Pranab Kumar Dan(IIT)

PREFACE

This book addresses Industrial engineering and Non Destructive Testing ,analysis in the context of an overall effort to achieve quality. It is designed for use as a primary handbook with updated example,as a supplementary text in a material/mechanical/Industrial engineering course, and as a resource for academic and practical approach.

The main characteristics of this book are:

- It presents modern optimization techniques suitable for near-term application, with sufficient technical background to understand their domain of applicability and to consider variations to suit technical and organizational constraints

- It presents NDT of techniques suitable for near-term application, with sufficient technical background to understand their domain of applicability and to consider variations to suit technical and organizational constraints.

CONTENTS

1 INTRODUCTION OF INDUSTRIAL ENGINEERING

- 1.1. Definition
- 1.2. Related developments of IE
- 1.3. Engineering Industry in Present India
- 1.4. **Engineering Industry in India**
- 1.5. Chronology of Industrial Engineering
- 1.6. Industrial Revolution
- 1.7. Significant Events Of Industrial Engineering
- 1.8. Objective Of Industrial Engineering
- 1.9. Responsibility Of Industrial Engineering
- 1.10. Techniques Of Industrial Engineering
- 1.11. System Engineering
- 1.12. Manufacturing system engineering
- 1.13. Introduction of CAD/CAM in IE

2 OPTIMIAZATION APPROCHES IN INDUSTRIAL ENGINEDERING

- 2.1 Optimization and Decision Theory
- 2.2 Introduction Of MCDM Methods
- 2.3 Present Approaches In MCDM Methods With Example

2.4   IE in mechanical design optimization.

3   CONCEPT OF APPLIED MANAGEMENT IN IE
3.1   Introduction of Quality management
3.2   Dimensions of Quality
3.3   Six sigma
3.4   ISO-9000 Definition
3.5   Typical focus areas in IE
3.6   Concept of Inventory Management

4   Industrial Engineering in NDT environment
4.1 Introduction of NDT
4.2 Methods of Inspection
4.3 Concept of inspection
4.4 Design Inspection Strategy-An practical example
4.5 IE approach in NDT environment-An Example

5   REFERENCE

INTRODUCTION OF INDUSTRIAL ENGINEERING

It is best to say that an Industrial Engineer should be an all-rounder. He/She should be able to take up any work assignment in his capacity. Indeed, there are a few sets of skills that a person must possess or instill in with time.

Complex Problem Solving

Reading Comprehension

Knowledge of scientific way of management

Good negotiating and interpersonal skills

Strong organizational skills

Good numeracy skills and knowledge of statistics

A high standard of computer literacy

Business strategy, operations and documentation skills

Having an organized vision, strategic objectives and a long-range plan

1.1 Definition

Industrial Engineering is concerned with the design, improvement and installation of integrated system of people, materials and equipment. It's utilizing and coordinating humans, machines, and materials to attain a desired output rate with the optimum utilization of energy, knowledge, money, and time.

The following formal definition of industrial engineering has been adopted by the Institute of Industrial Engineers (IIE):

*"**Industrial Engineering** is concerned with the design, improvement, and installation of integrated systems of people, materials, information, equipment and energy. It draws upon specialized knowledge and skill in the mathematical, physical, and social sciences together with the principles and methods of engineering analysis and design to specify, predict, and evaluate the results to be obtained from such system".*

Scope

The extent of industrial engineering is evidenced by the wide range of such activities as research in biotechnology, development of new concepts of information processing, design of automated factories, and operation of incentive wage plans.

Diversity

Industrial engineering is a diverse discipline concerned with the design, improvement, installation, and management of integrated systems of people, materials, and equipment for all kinds of manufacturing and service operations.

IE is concerned with performance measures and standards, research of new products and product applications, ways to improve use of scarce resources and many other problem solving adventures.
IE draws upon specialized knowledge and skill in the mathematical, physical, and social sciences, together with a strong background in engineering analysis and design and the management sciences to specify, predict, and evaluate the performance from such systems.

Employment

An Industrial Engineer may be employed in almost any type of industry, business or institution, from retail establishments to manufacturing plants to government offices to hospitals.

1.2 Related developments affected by Industrial engineering

1- Impact of Operations Research

The development of industrial engineering has been greatly influenced by the impact of an analysis approach called operations research.

This approach originated in England and the United States during 2nd World War and was aimed at solving difficult war-related problems through the use of science, mathematics, behavioral science, probability theory, and statistics.

2. Impact of Digital Computers

Digital computers permit the rapid and accurate handling of vast quantities of data, thereby permitting the IE to design systems for effectively managing and controlling large, complex operations.

The digital computer also permits the IE to construct computer simulation models of manufacturing facilities and the like in order to evaluate the effectiveness of alternative facility configurations, different management policies, and other management considerations.

Computer simulation is emerging as the most widely used IE technique. The development and widespread utilization of personal computers is having an exciting impact on the practice of industrial engineering.

3. Emergence of Service Industries

In the early days of the industrial engineering profession, IE practice was applied almost fully in manufacturing organizations. After the 2nd World War there was a growing awareness that the principles and techniques of IE were also applicable in non-manufacturing environments.

Thousands of Industrial Engineers are employed by government organizations to increase efficiency, reduce paperwork, design computerized management control systems, implement project management techniques, monitor the quality and reliability of vendor-supplied purchases, and for many other functions

1.3 Industrial Engineering in India

Industrial management and production engineering is been highlighted as a very specialized branch developed for serving the industries. On one hand we take production engineering which is an amalgamation of manufacturing technology and science. While on the other hand we have industrial engineering that is specialized for industrial processes, operations and management. So being a production and industrial engineer we could simply serve as an asset for the industries. Scope of IPE is very vast and is preferred by every organization because industrial engineers are employed to manage and optimize resources and operations. After taking a b.tech qualification in IPE we could pursue specialization in industrial engineering from IITs or NITIE (national institute of industrial engineering, Mumbai). This provides a great platform and a great future.

Normally, industrial engineers are hired to increase the productivity of a manufacturing facility while reducing wasted materials.

Many businesses consult with industrial engineers before they open a new factory or plant, because industrial engineers are known to be good at logistics, they can help businesses identify the most cost effective location for their production facilities.

An industrial engineer:

- Develops and evaluates pilot programs, simulations, and prototypes, and implement new and enhanced processes and tools,,Assist in financial planning and cost analysis
- Improve ways to distribute goods and services and determine most viable plant and factory locations, May have to work in non-manufacturing areas, they sometimes help employers figure out how they should evaluate their employees and how much they should be paid.
- Study product requirements and design, manufacturing and information systems to meet requirements
- Evaluate accuracy of production and testing equipment.
- Some job titles that are offered under Industrial Engineering

They have been employed as a Ergonomist, Plant Engineer, Process Engineers, Quality Engineers, Industrial Managers, Operations Analyst, Manufacturing Engineer, Quality Control Engineer

1.4 Engineering Industry in India

India is on the threshold of major reforms and is poised to become the third-largest economy of the world by 2030 cause- This is according to a report by the UK think tank Centre for Economics Business and Research (CEBR). We have an increasingly tech-savvy, educated population, consisting of skilled and intelligent labour and you. Of course is part of this. Which means you are at the cusp of making a great career for yourself as well as contribute to the glorious development that this country. This is especially true for students aiming to go for engineering courses. If you have the right passion, the future is bright.

There is opportunities and scope in almost every field although the scope might keep changing as new trends, technologies and requirements emerge with each year. The key question that you've to ask yourself before choosing an engineering branch is whether you are passionate in that particular field. If you are choosing a course simply based on its job possibilities, then you are going to be stuck in a course that you hate and you have to struggle to complete it. Then you will be trapped in a job that you have no skill or interest in. So it is very important to go for a branch that is of your interest. Here we are listed the top 10 engineering branches that you can opt for based on your interests and skills.

Inflow of foreign investments

- Cumulative FDI inflows increased to US$ 28.22 billion in FY16 from US$ 8.9 billion in FY10.
- The government's increasing focus on attracting foreign investors in manufacturing and infrastructure is likely to boost FDI in the sector.

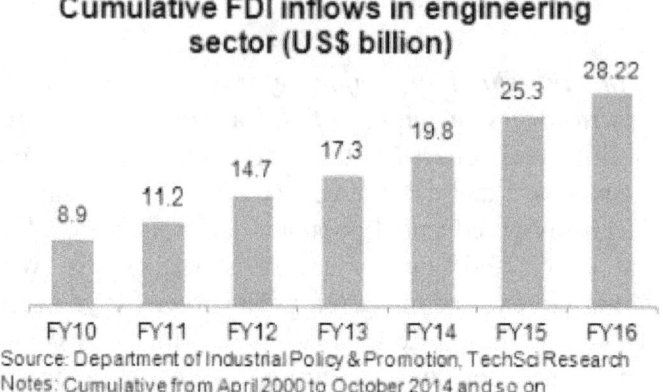

Source: Department of Industrial Policy & Promotion, TechSci Research
Notes: Cumulative from April 2000 to October 2014 and so on

Introduction

The Indian Engineering sector has witnessed a remarkable growth over the last few years driven by increased investments in infrastructure and industrial production. The engineering sector, being closely associated with the manufacturing and infrastructure sectors, is of strategic importance to India's economy.

India on its quest to become a global superpower has made significant strides towards the development of its engineering sector. The Government of India has appointed the Engineering Export Promotion Council (EEPC) as the apex body in charge of promotion of engineering goods, products and services from India. India exports transport equipment, capital goods, other machinery/equipment and light engineering products such as castings, forgings and fasteners to various countries of the world.

India became a permanent member of the Washington Accord (WA) in June 2014. The country is now a part of an exclusive group of 17 countries who are permanent signatories of the WA, an elite international agreement on engineering studies and mobility of engineers.

Market size

The capital goods & engineering turnover in India is expected to reach US$ 125.4 billion by FY17.

India exports its engineering goods mostly to the US and Europe, which accounts for over 60 per cent of the total exports. Recently, India's engineering exports to Japan and South Korea have also increased with shipments to these two countries rising by 16 and 60 per cent respectively. Sri Lanka, Nepal and Bangladesh have also emerged as the major destinations for India's engineering exports.

Engineering exports from India increased for the sixth straight month at 12.4 per cent year-on-year to US$ 5.3 billion in January 2017, outperforming that of the overall merchandise export.

Investments

The engineering sector in India attracts immense interest from foreign players as it enjoys a comparative advantage in terms of manufacturing costs, technology and innovation. The above, coupled with favorable regulatory policies and growth in the manufacturing sector has enabled several foreign players to invest in India.

The foreign direct investment (FDI) inflows into India's miscellaneous mechanical and engineering industries during April 2000 to December 2016 stood at around US$ 3,296.07 million, as per data released by the Department of Industries Policy and Promotion (DIPP).

In the recent past there have been many major investments and developments in the Indian engineering and design sector:

- Engineers India Ltd and Gazprom PJSC, the respective domestic companies of India and Russia in the engineering and oil and gas sectors, will prepare a blueprint for laying a gas pipeline between India and Russia, which is expected to help India diversify its energy mix and increase trade with Russia.
- Hexagon Capability Centre India (HCCI) in collaboration with National Institute of Technology Karnataka (NITK), Surathkal, launched first-of-its-kind NextGen 3D Lab costing Rs 7.7 crore (US$ 1.15 million) at NITK Campus.

The lab aims at making budding engineers industry-ready by the time they graduate.
- Engineering and construction major L&T entered into a joint venture with European defence major Matra BAE Dynamics Alenia (MBDA) Missile Systems for development of missiles in India. L&T will own 51 per cent stake in the JV named L&T MBDA Missile Systems and the rest 49 with the European partner.
- American plane maker Boeing Corporation has launched the Boeing India Engineering & Technology Center in Bengaluru. The centre will employ hundreds of locals who will work to support Boeing, including its information technology & data analytics, engineering, research and technology, and tests.
- Reliance Defence and Engineering Ltd said it has signed an agreement with the US Navy for undertaking service, maintenance and repair of Seventh Fleet of US Navy at the Reliance Shipyard at Pipavav in Gujarat.
- Rolta, an Information Technology (IT), engineering and geospatial services provider, has been awarded a seven-year, multi-million pound contract by a UK based major utility company UK Power Networks, to manage and update the firm's spatially-enabled network asset information.
- India's Texmaco Rail & Engineering has signed a memorandum of understanding (MoU) with Russia's ROSOBORONEXPORT (ROE) for modernisation of Armoured Vehicles operated by the Indian Army.
- Volvo Penta, a marine and industrial power system manufacturer, plans to produce five and eight litre industrial engines at the VE Powertrain (VEPT) plant in Pithampura near Indore from 2017.
- Toshiba Transmission and Distribution Systems (India) Pvt Ltd has bagged Rs 226 crore (US$ 33.9 million) contract from Kenya Power and Lighting Company for around 8,000 distribution transformers.
- L&T Hydrocarbon Engineering (LTHE), a subsidiary of Larsen & Toubro, has bagged an onshore EPC contract of over Rs 650 crore (US$ 97.5 million) from Gujarat State Fertilisers and Chemicals (GSFC) for setting up 40,000

- million tonnes per annum (mtpa) Melamine Plant at Fertiliser Nagar, Vadodara.
- Toshiba Group's water services company UEM India bagged Rs 220 crore (US$ 33 million) design, builds and operate (DBO) contract for a wastewater treatment and recycling plant in Oman.
- Essar Projects, the engineering, procurement & construction (EPC) arm of Essar Group, in a joint venture with Italy's Saipem has won a US$ 1.57 billion contract from Kuwait National Petroleum Company (KNPC) for setting up part of the Al-Zour Refinery Project in Kuwait.
- India's engineering and construction major, Punj Lloyd, won an order worth Rs 477 crore (US$ 71.55 million) for Ennore LNG tankage project from Mitsubishi Heavy Industries of Japan.
- Honeywell Turbo Technologies partnered with Tata to develop their first ever petrol turbocharged engine. The new Tata Revotron 1.2T engine launched in the 2014 Tata Zest delivers improved power and torque and a multi-drive mode, according to a Honeywell statement. Honeywell's engineering teams in Pune and Bangalore leveraged local capabilities and global expertise in petrol turbo technologies to address the specific needs of a local customer.
- The engineering and R&D division of HCL Technologies will likely cross the US$ 1 billion mark in the next financial year as the company sees larger deals in a market that's widely expected to be the next big source of growth for the Indian IT sector.
- Engineers India Ltd (EIL) inked a US$ 139 million consultancy deal for a 20 million tonnes (MT) refinery and polypropylene plant being built in Nigeria by Dangote Group.
- Tractebel Engineering (India) acquired Cethar Consulting Engineers Ltd. (CCE), the renowned and respected engineering consultancy company. This acquisition makes Tractebel Engineering a key player in thermal tower sector in India and strongly enhances the portfolio of offerings, which include gas pipelines, Liquefied Natural Gas, hydro power sector.

- Bharat Forge acquired Mecanique Generate Langroise (MGL), French oil and gas machining company, via its German arm CDP Bharat Forge GmbH. Bharat Forge will benefit from MGL's expertise in precision machining and other high value processes like cladding which have critical application in the oil and gas industry.
- Leading aircraft maker Airbus announced it has begun sourcing components for almost all its jets from India and it aims to take its cumulative sourcing from India to US$ 2 billion by 2020.
- Larsen & Toubro bagged construction orders worth Rs 1,099 crore (US$ 164.85 million) which included jobs from power transmission and distribution sector worth Rs 517 crore (US$ 77.55 million) and a rural electrification project under the Rajiv Gandhi Grameen Vidyutikaran Yojana (RGGVY) scheme at Gorakhpur in Uttar Pradesh.

Government Initiatives

The Indian engineering sector is of strategic importance to the economy owing to its intense integration with other industry segments. The sector has been de-licensed and enjoys 100 per cent FDI. With the aim to boost the manufacturing sector, the government has relaxed the excise duties on factory gate tax, capital goods, consumer durables and vehicles.

- The Government of India is planning to merge 6 engineering consulting Public Sector Units (PSUs) to create a mega consultancy firm that can take up projects across sectors and compete with the likes of Bechtel of the US and domestic majors like Larsen & Toubro (L&T).
- Steps have also been taken to encourage companies to perform and grow better. For instance, EIL was recently conferred the Navaratna status after it fulfilled the criteria set by the Department of Public Enterprises, Ministry of Heavy Industries and Public Enterprises, Government of India. The conferred status would give the state-owned firm more financial and operational autonomy.
- Government of India has also taken initiatives to provide a level playing field to domestic and foreign private players bidding for the government contracts in defence sector. The

government has withdrawn excise and customs duty exemptions granted to goods manufactured and supplied to the defence ministry by state-owned defence firms. These steps will also encourage participation of foreign Original Equipment Manufacturers such as Boeing, Airbus, Lockheed Martin, BAE Systems, etc., in the sector.
- The Government of India and the World Bank have signed a US$ 201.50 million IDA credit agreement for the Third Technical Education Quality Improvement Programme (TEQIP III), aimed at improving the efficiency, quality and equity of engineering education across several focus states.
- Prime Minister, Mr Narendra Modi announced a partnership between Bloomberg Philanthropies and the Ministry of Urban Development, Government of India, to advance the "Smart Cities Initiative." The Smart Cities Initiative is a historic effort to promote economic growth, improve governance, and deliver more effective and efficient public services to India's urban residents.

Road Ahead

The engineering sector is a growing market. Spending on engineering services is projected to increase to US$ 1.1 trillion by 2020.

Exchange Rate Used: INR 1 = US$ 0.015 as on February 9, 2017

References: *Media reports, Press releases, EEPC India, Press Information Bureau (PIB), Department of Industrial Policy and Promotion (DIPP)*

1.5 Chronology of Industrial Engineering

- Engineering history lies back to the beginning of civilization. Until the end of 17hundreds, production meant crafts (A craftsman used to treat material and assemble the pieces). Until then a single person used to Plan, Select and supply material, Produce and control

- In 1776, James Watt invented the steam engine. (Turning steam power into mechanical power)

- Charles Babbage visited factories in England and the United States in the early 1800's and began a systematic recording of the details involved in many factory operations.

Frederick W. Taylor is credited with recognizing the potential improvements to be gained through analyzing the work content of a job and designing the job for maximum efficiency.

Frank B. Gilbreth extended Taylor's work considerably. Gilbreth's primary contribution was the identification, analysis and measurement of fundamental motions involved in performing work.

Another early pioneer in industrial engineering was Henry L. Gantt, who developed the so-called Gantt chart. The Gantt chart was a significant contribution in that it provided a systematic graphical procedure for pre-planning and scheduling work activities, reviewing progress, and updating the schedule. Gantt charts are still in widespread use today.

During the 1920s and 1930s much fundamental work was done on economic aspects of managerial decisions, inventory problems, incentive plans, factory layout problems, material handling problems, and principles of organization.

1.6 Industrial Revolution

The Industrial Revolution was the transition to new manufacturing processes in the period from about 1760 to sometime between 1820 and 1840. This transition included going from hand production methods to machines, new chemical manufacturing and iron production processes, improved efficiency of water power, the increasing use of steam power, the development of machine tools and the rise of the factory system. Textiles were the dominant industry of the Industrial Revolution in terms of employment, value of output and capital invested; the textile industry was also the first to use modern production methods.

The Industrial Revolution created a demand for metal parts used in machinery. This led to the development of several machine tools for cutting metal parts. They have their origins in the tools developed in the 18th century by makers of clocks and watches and scientific instrument makers to enable them to batch-produce small mechanisms.

Before the advent of machine tools, metal was worked manually using the basic hand tools of hammers, files, scrapers, saws and chisels. Consequently, the use of metal was kept to a minimum. Wooden components had the disadvantage of changing dimensions with temperature and humidity, and the various joints tended to rack (work loose) over time. As the Industrial Revolution progressed, machines with metal parts and frames became more common. Hand methods of production were very laborious and costly and precision was difficult to achieve. Pre-industrial machinery was built by various craftsmen – millwrights built water and wind mills, carpenters made wooden framing, and smiths and turners made metal parts.

Source:Internet

1.7 Significant Events Of Industrial Engineering

The Industrial Revolution marks a major turning point in history; almost every aspect of daily life was influenced in some way. In particular, average income and population began to exhibit unprecedented sustained growth. Some economists say that the major impact of the Industrial Revolution was that the standard of living for the general population began to increase consistently for the first time in history.

➢ Division of labor (Smith, 1776)
➢ Standardized parts (Whitney, 1800)
➢ Scientific management (Taylor, 1881)
➢ Coordinated assembly line (Ford 1913)
➢ Gantt charts (Gantt, 1916)

- Motion study (the Gilbreths, 1922)
- Quality control (Shewhart, 1924)
- CPM/PERT (Dupont, 1957)
- MRP (Orlicky, 1960)
- CAD
- Flexible manufacturing systems (FMS)
- Computer integrated manufacturing (CIM)

1.8 Objective Of Industrial Engineering

- Design facilities, management systems, operating procedures.
- Improve planning and allocation of scarce resources. Enhance plant environment and quality of people's working life.
- Evaluate reliability and quality performance.
- Develop management control systems to aid in financial planning and cost analysis.
- Implement office systems, procedures, and policies.
- Analyze complex business problems by operations research.
- Conduct organization studies, plant location surveys, and system effectiveness studies.
- Study potential markets for goods and services, raw material sources, labor supply, energy resources, financing, and taxes.
- Develop applications of new processing, automation, and control technology.
- Install data processing, management information, wage incentive systems.
- Develop performance standards, job evaluation, and wage and salary programs.
- Research new products and product applications.
- Improve productivity through application of technology and human factors.
- Select operating processes and methods to do a task with proper tools and equipment.
And so on.

1.9 Responsibility Of Industrial Engineering

Industrial Engineering is concerned with the design, improvement and installation of integrated system of people, materials and equipment. It draws upon the specialized knowledge and skill in mathematical, physical and social science together with the principles and methods of engineering analysis to predict and evaluate the results to be obtained from such systems.

Industrial Engineering is unique in the sense that it is a link between technology and Management, technology and economics and technology and science. Industrial Engineering discipline is different from other engineering disciplines. It can be stated that Industrial Engineering deals with people as well as things.

- To understand the role of industrial engineering (IE) it is helpful to learn the historical developments that were involved in the development of IE.
- Principles of early engineering were first taught in military academies and were concerned primarily with road and bridge construction and with defenses.

1.10 Techniques of Industrial Engineering

IEs Work in Many Types of Industries
- Aerospace & Airplanes
- Aluminum & Steel
- Banking
- Ceramics
- Construction
- Consulting
- Electronics Assembly
- Energy
- Entertainment
- Forestry & Logging

- Insurance
- Materials Testing
- Medical Services
- Military
- Mining
- Oil & Gas
- Plastics & Forming
- Retail
- Shipbuilding
- State & Federal Government
- Transportation and so on.

Some Techniques Utilized by IEs
- Benchmarking
- Design of Experiments
- Employee Involvement
- Equipment Utilization
- Flow Diagramming
- Information & Data Flow Diagramming
- Interviewing for Information
- Lean Manufacturing
- Modeling & Testing
- Operations Auditing
- Organizational Analysis
- Pilot Programs
- Plant & Equipment Layout
- Project Management
- Simulation
- Six Sigma projects
- Statistical Analysis
- Strategic Planning
- Theory of Constraints
- Time Studies
- Work Sampling

1.11 System Engineering

Systems engineering signifies only an approach and, more recently, a discipline in engineering. The aim of education in systems engineering is to formalize various approaches simply and in doing so, identify new methods and research opportunities similar to that which occurs in other fields of engineering. As an approach, systems engineering is holistic and interdisciplinary in flavour.

The term *systems engineering* can be traced back to Bell Telephone Laboratories in the 1940s.[1] The need to identify and manipulate the properties of a system as a whole, which in complex engineering projects may greatly differ from the sum of the parts' properties, motivated various industries, especially those developing systems for the U.S. Military, to apply the discipline.

In 1990, a professional society for systems engineering, the National Council on Systems Engineering (NCOSE), was founded by representatives from a number of U.S. corporations and organizations. NCOSE was created to address the need for improvements in systems engineering practices and education. As a result of growing involvement from systems engineers outside of the U.S., the name of the organization was changed to the International Council on Systems Engineering (INCOSE) in 1995.[4] Schools in several countries offer graduate programs in systems engineering, and continuing education options are also available for practicing engineers

Systems engineering focuses on analyzing and eliciting customer needs and required functionality early in the development cycle, documenting requirements, then proceeding with design synthesis and system validation while considering the complete problem, the system lifecycle. This includes fully understanding all of the stakeholders involved. Oliver et al. claim that the systems engineering process can be decomposed into

- a Systems Engineering Technical Process, and
- a Systems Engineering Management Process.

Quality function deployment (QFD)

Quality function deployment (QFD) is the translation of user requirements and requests into product designs. The goal of QFD is to build a product that does exactly what the customer wants instead of delivering a product that emphasizes expertise the builder already has.

he concept of QFD, which is a combination of the ideas of Deming, Juran and Taguchi, was introduced in Japan by Yoji Akao in 1966. The actual Japanese name for the methodology is *hin shitsu, ki nou, ten kai*. Translating into English is difficult since the terms have multiple meanings. The best translation is Quality Function Deployment (see figure 1). QFD is designed to help organizations understand the customer's requirements, prioritize them, and respond to them in an organized way. QFD is a way of listening to the customer. Using QFD, system developers are forced to answer the following three questions: 1) What are the qualities the customer desires from the product, 2) what must the product do (what are the main functions of the product), and 3) how can we as the developer, considering resources available, best provide what the customer wants?

Figure 1. QFD name in Japanese (Guinta and Praizler, 1993, p. 5)

QFD in Systems Engineering

A concise definition of systems engineering is needed as well as a framework, or model, with which to show the incorporation of the elements of QFD. There are many definitions of systems engineering, but Blanchard (1991) provides an interesting description of systems engineering that incorporates a central concept of QFD. He states that system engineering is not an "engineering discipline" in itself, but " ... system engineering involves the efforts pertaining to the overall design process employed in the evolution of a system from the point when a need is first identified, through production and/or construction and the ultimate distribution of that system for consumer use. The objective is to meet the requirements of the consumer in an effective and efficient manner." Blanchard makes a point of identifying the role the customer plays in the overall engineering effort. The ultimate goal is to satisfy the customer. System engineering by itself is a type-specific methodology used in the development of products, and incorporates a complete requirements and design methodology. System engineering emphasizes completeness; while QFD emphasizes focus (Bicknell and Bicknell, 1993).

CHAPTER-2

OPTIMIAZATION APPROCHES IN INDUSTRIAL ENGINEDERING

2.1 Introduction of Optimization and Decision Theory

It is hard to imagine any science or engineering discipline, where optimization and/or decision theoretical models are not of major concern. Telecommunication systems, engineering design, logistics networks, manufacturing plants, biological and financial systems - to name a fewrely heavily on optimization approaches. For instance, an important design problem in structural mechanics is solved by finding the optimal solution of the corresponding equilibrium problem. Moreover, companies can reduce their costs significantly by applying optimization techniques in order to design and operate their supply chains.

The Optimization and Decision Theory research are mainly involves the development of algorithms based on linear, nonlinear, integer, dynamic and stochastic programming techniques. Frequently, optimal algorithms require excessive computational resources and time when applied to major problems encountered in real life. In such cases, it is crucial to develop effective heuristic algorithms by exploiting the problem structure. Hence, research in this Group employs both classical heuristics and metaheuristics, which have become increasingly popular recently.

The set of problems studied in the group includes, but is not restricted to: machine tool selection, energy efficient routing in wireless sensor and ad hoc networks, global optimization, constrained optimization and project scheduling.

If in machine designing, the material and geometrical parameters are optimized simultaneously then it is common to assume empirical formulas approximating a relation between material parameters for example the bending fatigue limit (Sbf) and ultimate tensile strength (UTS) as a function of hardness. If the choice of material is limited to a list of pre-defined candidates, then two difficulties can be appeared. First, a discrete optimization process should be followed against material parameters. Second, properties of different alternatives materials may not indicate any obvious correlation in the given list. The main goal is to choose material with best characteristic among alternatives.

2.2 Introduction Of MCDM Methods

Multiple criteria decision making (MCDM) is the process of selecting the best alternative from a set of feasible alternatives considering multiple conflicting criteria. In precise terms criteria are considered to be 'strictly' conflicting if the increase in satisfaction of one results in a decrease in satisfaction of the other. An MCDM process always contains at least two alternatives and two conflicting criteria (Bhattacharya et al., 2003). MCDM are divided two broad categories: Multiple Attribute Decision Making (MADM) and Multiple Objective Decision Making (MODM). Several useful tools for solving of MCDM problems are

- Simple Additive Weighting method (SAW)
- Technique for Order Preference by Similarity to Ideal Solution (TOPSIS)
- Multi Objective Optimization Ratio Analysis(MOORA)
- Analytical Hierarchy Method (AHP)
- Analytical Network Method ANP etc.
  - Here discussed some example tool of MCDM theory
  - **SIMPLE ADDITIVE WEIGHTING (SAW)**

Step 1 Formation of decision matrix: Criterion outcomes of decision alternatives can be collected in a table called Decision Matrix comprised of a set of columns and rows. The matrix rows represent decision alternatives, with matrix columns representing criteria. A value found at the intersection of row and column in the matrix represents a criterion outcome - a measured or predicted performance of a decision alternative on a criterion. The decision matrix is a central structure of the MCDA/MCDM since it contains the data for comparison of decision alternatives.

$$X = \begin{array}{c} \\ A_1 \\ \vdots \\ A_i \\ \vdots \\ A_m \end{array} \begin{array}{cccccc} C_1 & & C_j & & C_n \\ \begin{bmatrix} x_{11} & \cdots & x_{1j} & \cdots & x_{1n} \\ \vdots & \cdots & \vdots & \cdots & \vdots \\ x_{i1} & \cdots & x_{ij} & \cdots & x_{in} \\ \vdots & \cdots & \vdots & \cdots & \vdots \\ x_{m1} & \cdots & x_{mj} & \cdots & x_{mn} \end{bmatrix} \end{array}$$

(1)

x_{ij} is the performance rating of alternative i with respect to criterion j,

A_j is i^{th} alternative, C_j is the jth criterion

Step 2 Formation of Weight Matrix

Different importance weights to various criteria may be awarded by the decision makers. These importance weights forms the weight as follows.

$$W=[W_1 \cdots W_j \cdots W_n]$$

(2)

Step 3 Normalization of performance rating

Units and dimensions of performance ratings of columns under criteria differ. For the purpose of comparison, these performance ratings are converted into dimensionless units by normalization using following equations

$$\bar{x}_{ij} = \frac{x_{ij}}{\max_i(x_{ij})} \text{ for benefit criteria } j$$

(3)

$$\bar{x}_{ij} = \frac{\min_i(x_{ij})}{x_{ij}} \text{ for non-benefit criteria } j$$

(4)

Normalized decision matrix

$$\overline{X} = \begin{matrix} A_1 \\ A_2 \\ \vdots \\ A_m \end{matrix} \begin{bmatrix} \overline{x}_{11} & \cdots & \overline{x}_{1j} & \cdots & \overline{x}_{1n} \\ \vdots & & \vdots & & \vdots \\ \overline{x}_{i1} & \cdots & \overline{x}_{ij} & \cdots & \overline{x}_{in} \\ \vdots & & \vdots & & \vdots \\ \overline{x}_{m1} & & \overline{x}_{mj} & & \overline{x}_{mn} \end{bmatrix}_{m \times n}$$

(5)

Step 4 composite score: Computation of composite score (CS_i) for alternative i

$$CS_i = \sum_{j=1}^{n} \left(\overline{w}_j * \overline{x}_{ij} \right)$$

Step 5 Ranking and selection of best alternative: Ranking of products in descending order of composite scores (CS_i).

> **Technique for Order Preference by Similarity to Ideal Solution (TOPSIS)**

TOPSIS is an evaluation method that is often used to solve MCDM problems [2, 3].

It has a number of applications [4, 5] in practice, such as comparison of company performances, financial ratio performance within a specific industry and financial investment in advanced manufacturing systems, etc. However, there are also some limits to it. So far, the work on how to improve original TOPSIS method has mainly emphasized on improving the weight to sensitize the R value [6, 7]. Besides, there has also been improvement on formula of the R value, such as the 'Miqiezhi' method [8]. Because of the complexity of evaluation problems, a better and simpler method is required to understand the inherent relationship between the R value and alternative evaluation. In this report, a novel, modified TOPSIS (M-TOPSIS) method is described as a process of calculating the distance between the alternatives and the reference points in the D+ D--plane and constructing the R value to evaluate quality of alternative.

✦ Algorithm of TOPSIS method under MCDM

The idea of TOPSIS can be expressed in a series of steps:

Step1 All the original criteria receive tendency treatment. We usually transform the cost criteria into benefit criteria, which is shown in detail as follows ;

(i) The reciprocal ratio method ($X_{ij} = 1/X_{ij}$), refers to the absolute criteria;

(ii) The difference method ($X_{ij} = 1 - X_{ij}$), refers to the relative criteria.

After tendency treatment, construct a matrix

$$X' = [X'_{ij}]_{n \times m}, i = 1, 2 \ldots, n; \quad j = 1, 2 \ldots, m.$$

Step2 Calculate the normalized decision matrix A. The normalized value aij is calculated as

$$A = [a_{ij}]_{n \times m}, \ a_{ij} = X'_{ij} / \sqrt{\sum_{i=1}^{n} (X'_{ij})^2} \quad i = 1, 2 \ldots, n; \quad j = 1, 2 \ldots, m. \qquad (2.2)$$

Step3 Determine the positive ideal and negative ideal solution from the matrix A.

$$A^+ = (a_{i1}^+, a_{i2}^+, \ldots, a_{im}^+), a_{ij}^+ = \max_{1 \leqslant i \leqslant n}(a_{ij}), \quad j = 1, 2 \ldots, m$$

$$A^- = (a_{i1}^-, a_{i2}^-, \ldots, a_{im}^-), a_{ij}^- = \min_{1 \leqslant i \leqslant n}(a_{ij}), \quad j = 1, 2 \ldots, m$$

Step4 Calculate the separation measures, using the n-dimensional Euclidean distance. The separation of each alternative from the positive ideal solution is given as:

$$D_i^+ = \sqrt{\sum_{j=1}^{m} W_j (a_{ij}^+ - a_{ij})^2}$$

Similarly, the separation from the negative ideal solution is given as

$$D_i^- = \sqrt{\sum_{j=1}^{m} W_j(a_{ij}^- - a_{ij})^2}$$

Step 5 For each alternative, calculate the ratio Ri as:

$$R_i = \frac{D_i^-}{D_i^- + D_i^+} \quad i = 1, 2 \ldots, n$$

Step 6 Rank alternatives in increasing order according to the ratio value of Ri in

step 5.

> ## MULTI OBJECTIVE OPTIMIZATION RATIO ANALYSIS (MOORA)

The MOORA method which was introduced by Brauers (Brauers, 2006) is such a multi objective optimization technique that can be successfully applied to solve various types of MCDM problems.

- Algorithm of MOORA method under MCDM

The MOORA method starts with a matrix of responses (performance measures) of different alternatives on different criteria (objectives or attributes). The matrix is shown below (Equation 1).

$$X = \begin{array}{c} \\ A_1 \\ \vdots \\ A_i \\ \vdots \\ A_m \end{array} \begin{array}{cccccc} C_1 & \cdots & C_j & \cdots & C_n \end{array} \\ \left[\begin{array}{ccccc} x_{11} & \cdots & x_{1j} & \cdots & x_{1n} \\ \vdots & \cdots & \vdots & \cdots & \vdots \\ x_{i1} & \cdots & x_{ij} & \cdots & x_{in} \\ \vdots & \cdots & \vdots & \cdots & \vdots \\ x_{m1} & \cdots & x_{mj} & \cdots & x_{mn} \end{array} \right]$$

(6)

Where x_{ij} is the performance rating (response) to the ith alternative (A_i) under jth criterion (C_j). m is the number of alternatives and n is the number of criteria.

The MOORA method employs a ratio system in which each response of an alternative on an attribute (criterion) is compared to a denominator. The denominator is a representative for all alternatives concerning that attribute (Brauers et al. 2007; Kalibatas and Turskis, 2008).

Brauers et al. (2008) considered various ratios such as the square root of the sum of squares of each alternative per objective, total ratios, Scharlig ratios, Weitendorf ratios, Jutter ratios, Stop ratios, Van Delft and Nijkamp ratios of maximum value, Korth ratios, Peldschus et al. and Peldschus ratios for nonlinear normalization. They concluded that the square root of the sum of squares of each alternative per objective is the best one for the denominator which is given below.

$$x_{ij}^* = \frac{x_{ij}}{\sqrt{\sum_{i=1}^{m}(x_{ij}^2)}}$$

(7)

x_{ij}^* is normalized value of response i with respect to attribute j. In the current research work, the maximum score under each attribute has also been used as the denominator of the ratio system and an effort has been made to exhibit that this ratio system is also suitable for finding the optimal solution. The following ratio system is the second best for normalization process in MOORA.

$$x_{ij}^* = \frac{x_{ij}}{\max_i(x_{ij})}$$

(8)

For the computation of normalized response using the above Eq. (2b), first the maximum score under each attribute is found. Then all the scores under certain attribute irrespective of benefit or non-benefit are divided by the concerned maximum score using Eq. (2b). x_{ij}^* is a dimensionless quantity in the interval [0,1] representing the normalized score of alternative *i* on attribute *j*. However, sometimes the interval could be [-1; 1]. For example in the case of productivity growth of some factories, industries, sectors, regions or countries may be negative instead of positive thus the interval becomes [-1;1] (Brauers *et al.*, 2008).

For multi-objective optimization these normalized performances are added in case of maximization and subtracted in case of minimization. Then the optimization problem becomes

$$y_i^* = \sum_{j=1}^{g} x_{ij}^* - \sum_{j=g+1}^{n} x_{ij}^*$$

(9)

Where *g* is the number of benefit criteria to be maximized and *(n-g)* is the number of non-benefit criteria to be minimized. y_i^* is final score of ith alternative with respect to all the attributes. In the above case it is assumed that all the attributes are of same importance.

$$y_i^* = \sum_{j=1}^{g} w_j * x_{ij}^* - \sum_{j=g+1}^{n} w_j * x_{ij}^*$$

(10)

Where w_j^* is the weight of jth attribute (criterion), which can be evaluated using any well-known approach either AHP or Entropy method. The value of y_i^* may be positive, negative or zero. These

y_i^* values are arranged in descending order. The best alternative is one which is associated with highest y_i^* value and the worst alternative is one which is associated with the lowest y_i^* value.

➤ ENTROPY

Entropy was originally a thermodynamic concept, first introduced into information theory by Shannon (see Shannon, 1948 [21]). It has been widely used in the engineering, socioeconomic and other fields. According to the basic principles of information theory, information is a measure of system's ordered degree, and the entropy is a measure of system's disorder degree.

Step1 Calculate p_{ij} (the ith scheme's jth indicator value's proportion).

$$p_{ij} = r_{ij} / \sum_{j=1}^{m} r_{ij}, \quad r_{ij}$$ is the ith scheme's jth indicator value.

Step2 Calculate the jth indicator's entropy value

$$e_j. e_j = -k \sum_{i=1}^{m} p_{ij} \ln p_{ij}, k = 1/\ln m,$$ m is the number of assessment schemes.

Step3 Calculate weight wj (jth indicator's weight).

$$w_j = (1-e_j)/\sum_{j=1}^{n}(1-e_j),$$ n is the number of indicators, and $0 \leq w_j \leq 1, \sum_{j=1}^{n} w_j = 1.$

In entropy method, the smaller the indicator's entropy value ej is, the bigger the variation extent of assessment value of indicators is, the more the amount of information provided, the greater the role of the indicator in the comprehensive evaluation, the higher its weight should be.

2.3 Present Approaches In MCDM Methods With Example

Optimal design of gear or any other machine requires the consideration of the two type parameters known as material and geometrical parameters. The choice of stronger material parameters may allow the choice of better geometrical parameters and vice versa. Very important difference among these two parameters is that the geometrical parameters are often varied independently. On the other hand, material parameters can be inherently correlated to each other and may not be varied independently. An example of which being the variation of the bending fatigue limit (Sbf) with the core hardness (HB) for some steel materials. If these parameters would be varied independently in an optimization case, it may result in infeasible solutions. Therefore, the final choice of material may not be possible within available data base.

MATERIAL PERFORMANCE INDICES

The main characteristics considered in the design of gears are:
- surface fatigue limit (Ssf),
- root bending fatigue limit (Sbf),
- wear resistance of tooth's flank
- High tensile strength to prevent failure against static loads
- High endurance strength to withstand dynamic loads
- Low coefficient of friction
- Good manufacturability

Generally cast iron, steel, brass and bronze are preferred for manufacturing metallic gears with cut teeth. Where smooth action is not important, cast iron gears with cut teeth may be employed. Commercially cut gears have a pitch line velocity of about 5 metre/second. For velocities larger than this, gear sets with non-metallic pinions as one member are used to eliminate vibration and noise. Non-metallic materials are made of various materials such as treated cotton pressed and moulded at high-pressure, synthetic resins of the phenol type and rawhide. Moisture affects rawhide pinions. Gears made of phenolic resins are self-supporting on the other hand other two types are supported by metal side plates at both ends of the plate. Large wheels are made with fretting rings to save alloy steels. Wheel centre is commonly cast from cast iron. The ring is forged or roll expanded from steel of the respective grade specified by the tooth design.

Problem Definition
An organization has got 9 different materials with different specifications for gear. The decision maker considered 7 selection criteria. The materials are as follows:

SL. NO.	Material	GRADE
Material 1	Cast iron	SAE J431-43500
Material 2	Ductile iron	EN-GJS 418
Material 3	S.G. iron	BS 2789
Material 4	Cast alloy steel	BS 2795
Material 5	Through hardened carbon steel	SAE 4140
Material 6	Surface hardened alloy steel	SAE 8620
Material 7	Carburised steel	SAE 8620
Material 8	Nitrided steel	EN40B
Material 9	Through hardened carbon steel	817M40

Table-01

The selection criteria are as follows:

C1	Surface Hardness (Bhn)
C2	Core Hardness (Bhn)
C3	Surface Fatigue Limit (MPa)
C4	Bending Fatigue Limit (MPa)
C5	UTS (MPa)
C6	Cost (INR) Per kg
C7	Supply Lead Time (In week)

Table-2

Out of 7 criteria, 5 criteria viz. C1: Surface Hardness (Bhn), C2: Core Hardness (Bhn), C3: Surface Fatigue Limit (MPa), C4: Bending Fatigue Limit (MPa), C5: UTS (MPa) are beneficial criteria because their higher values are desirable and remaining viz. C6: Cost (INR) Per kg, C7: Supply Lead Time(In week) are non-beneficial criteria because their lower values are desirable.

The objective of the decision maker is to assess the performance of the materials. Counseling the above 7 criteria to ultimately select the best material. The decision maker applied SAW, TOPSIS and MOORA methods for their simplicity, adaptability, applicability and is of applications. The decision matrix for the materials with respect to the criteria shown below:

Formation of decision matrix

MATERIAL	Grade	Surface (Bhn) Hardness (C1)	Core (Bhn) Hardness (C2)	Surface Fatigue Limit (MPa) (C3)	Bending Fatigue Limit (MPa) (C4)	UTS (MPa) (C5)	Cost (INR) Per kg (C6)	Supply Lead Time (In week) (C7)
Cast iron (M1)	SAE J431-43500	200	200	330	100	380	55	2
Ductile iron (M2)	EN-GJS 418	220	220	460	360	880	55	2
S.G. iron (M3)	BS 2789	240	240	550	340	845	47	3
Cast alloy steel (M4)	BS 2795	270	270	630	435	845	66	4
Through hardened carbon steel (M5)	SAE 4140	270	270	670	430	620	58	5
Surface hardened	SAE 8620	542	229	1160	680	1850	60	6

alloy steel (M6)								
Carburised steel (M7)	SAE 8620	647	297	1500	920	2300	60	5
Nitrided steel (M8)	EN40B	693	297	1250	760	1250	72	5
Through hardened carbon steel (M9)	817M40	185	185	500	430	635	74	5

Table: Suggested materials and their properties in a gear material selection problem[A]

Table-3

[A]Data(except material grade,cost and supply lead time) are taken form Hofmann (1990) where Vickers hardness values have been converted to Brinell values using conversion tables in http://www.gordonengland.co.uk/hardness/brinell_conversion_chart.htm

[A]Data (material grade,cost and supply lead time) are taken form Bill Forge Private Limited (Plant I)9C, Bommasandra Industrial Area,Hosur Road,Bangalore - 562 158,India

MATLAB

MATLAB supports a variety of graphs that enable you to present information effectively. The type of graph you select depends, to a large extent, on the nature of your data. The following list can help you select the appropriate graph:

- ✓ Bar and area graphs are useful to view results over time, comparing results, and displaying individual contribution to a total amount.
- ✓ Pie charts show individual contribution to a total amount.
- ✓ Histograms show of data values.
- ✓ Stem and stair step plots display discrete data.

ENTROPY METHOD							
criteria	C_1	C_2	C_3	C_4	C_5	C_6	C_7
weighted values	0.1635	0.1129	0.1634	0.1290	0.1143	0.1336	0.1833

- ✓ Compass, feather, and quiver plots display direction and velocity vectors.
- ✓ Contour plots show equivalued regions in data.
- ✓ Interactive plotting enables you to select data points to plot with the pointer.
  Animations add an addition data dimension by sequencing plots.

Computational result by MATLAB:

ENTROPY METHOD:

RESULT:

SAW METHOD									
Material	M_1	M_2	M_3	M_4	M_5	M_6	M_7	M_8	M_9
The values of (s)	3.3105	3.9933	3.9247	3.7710	4.0601	4.9866	6.1170	5.2557	3.0018
Arranging the final value in descending order:-				M7 > M8 > M6 > M5 > M2 > M3 > M4 > M1 > M9					

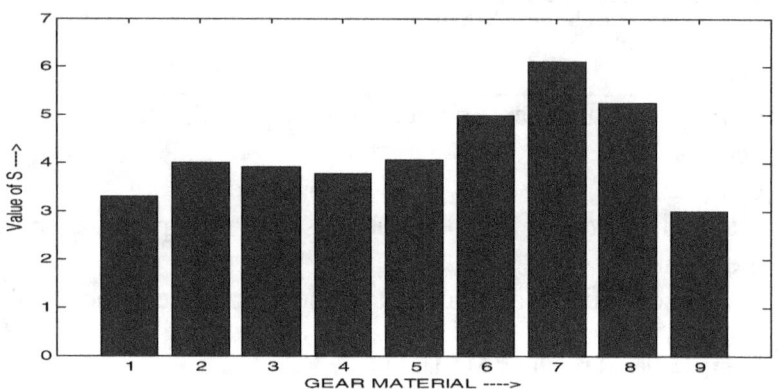

MOORA METHOD:

RESULT:

STEP 1   Determination of normalized decision matrix

	C_1	C_2	C_3	C_4	C_5	C_6	C_7
M_1	0.1623	0.2685	0.1258	0.0597	0.1000	0.2990	0.1538
M_2	0.1785	0.2953	0.1754	0.2149	0.2316	0.2990	0.1538
M_3	0.1948	0.3222	0.2097	0.2029	0.2224	0.2555	0.2308
M_4	0.2191	0.3625	0.2402	0.2596	0.2224	0.3588	0.3077
M_5	0.2191	0.3625	0.2555	0.3223	0.3131	0.3153	0.3846
M_6	0.4398	0.3074	0.4423	0.4058	0.4868	0.3262	0.4615
M_7	0.5250	0.3987	0.5720	0.5491	0.6052	0.3262	0.3846
M_8	0.5623	0.3987	0.4767	0.4536	0.3289	0.3914	0.3846
M_9	0.1501	0.2484	0.1907	0.2566	0.1671	0.4023	0.3846

STEP 2   Determination of weighted normalized decision matrix:

	C_1	C_2	C_3	C_4	C_5	C_6	C_7
M_1	0.0268	0.0301	0.0203	0.0075	0.0113	0.0409	0.0287
M_2	0.0295	0.0331	0.0283	0.0270	0.0261	0.0409	0.0287
M_3	0.0322	0.3222	0.2097	0.2029	0.2224	0.2555	0.2308
M_4	0.2191	0.3625	0.2402	0.2596	0.2224	0.3588	0.3077
M_5	0.0362	0.0406	0.0413	0.0404	0.0353	0.0431	0.0717
M_6	0.0727	0.0344	0.0715	0.0509	0.0548	0.0446	0.0860
M_7	0.0868	0.0446	0.0924	0.0689	0.0682	0.0446	0.0717
M_8	0.0929	0.0446	0.0770	0.0569	0.0371	0.0535	0.0717
M_9	0.0248	0.0278	0.0308	0.0322	0.0188	0.0550	0.0717

STEP 3:   Determination of weighted multi objective optimization:

(the value of a is the sum of all weighted normalized values for all beneficial column)

Material	M_1	M_2	M_3	M_4	M_5	M_6	M_7	M_8	M_9
The values of (a)	0.0960	0.1439	0.1526	0.1732	0.1938	0.2843	0.3609	0.3085	0.1344

The value of b is sum of all weighted normalized values for all non-beneficial column

Material	M_1	M_2	M_3	M_4	M_5	M_6	M_7	M_8	M_9
The values of (b)	0.0696	0.0696	0.0779	0.1064	0.1148	0.1306	0.1163	0.1252	0.1267

Material	M_1	M_2	M_3	M_4	M_5	M_6	M_7	M_8	M_9
The values of (a-b)	0.0264	0.0744	0.0747	0.0668	0.0790	0.1537	0.2446	0.1833	0.0077
Arranging the final value in descending order:-				M7 > M8 > M6 > M5 > M3 > M2> M4 > M1 > M9					

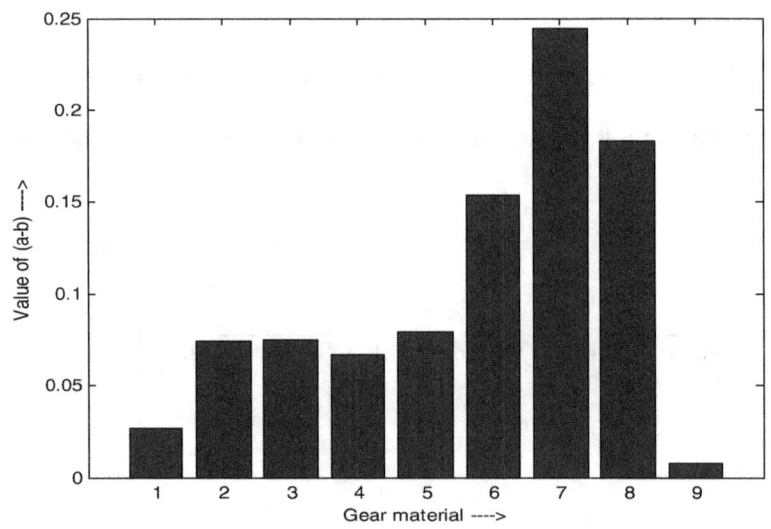

Fig:7

TOPSIS METHOD BY USING MATLAB:

Material	M_1	M_2	M_3	M_4	M_5	M_6	M_7	M_8	M_9
The values of R_i	0.3286	0.3944	0.3273	0.2967	0.3508	0.5560	0.6905	0.5941	0.1932
Arranging the final value in descending order:-					M7 > M8 > M6 > M2 > M5 > M1 > M3 > M4 > M9				

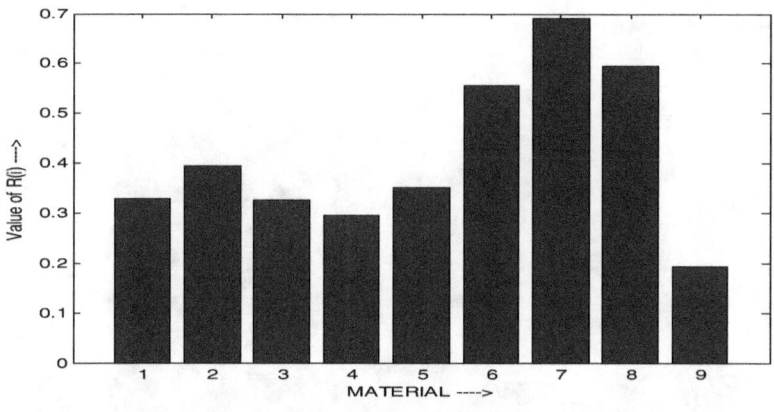

Fig:8

Comparative analysis of ranking of gear materials using MCDM methods:

MATERIAL	SAW (RANK)	MOORA (RANK)	TOPSIS (RANK)
M1	8	8	6
M2	5	6	4
M3	6	5	7
M4	7	7	8
M5	4	4	5

M6	3	3	3
M7	1	1	1
M8	2	2	2
M9	9	9	9

Table-4

DISCUSSION:

From the result we see that for the three different process of MCDM, the result is almost same. The ranking of 1st, 2ND, 3RD and 9th Materials are same for those three different processes. For the simplicity, prompt result getting the accurate value and also getting the best ranking we have used the MATLAB software. By this software we can also make rank of any system for any number of alternatives and criteria within a fraction of second with accuracy.

CONCLUSION

It is quite clear that selection of a proper Gear Materials for a given manufacturing application involves a large number of considerations. The use of SAW, TOPSIS and MOORA methods are observed to be quite capable and computationally easy to evaluate and select the proper material from a given set of alternatives. These methods use the measures of the considered criteria with their relative importance in order to arrive at the final ranking of the alternative Gear Materials. Thus, these popular MCDM methods can be successfully employed for solving any type of decision-making problems having any number of criteria and alternatives in the manufacturing domain. Use of MATLAB software makes MCDM problem simple and gives prompt results which is very essential in today's decision making environment.

2.4 IE in mechanical design optimization

Materials and process selection are key issues in optimal design of industrial products. Substituting and selecting materials for different machining parts is relatively common and often. Material selection is a difficult and subtle task, due to the immense number of different available materials. From this point of view paper deal with a set of major gear design criteria which are used for gear material selection. The paper introduces the decision models for best selective material processes. These models are explored in terms of fatigue theory as well as product life cycle is explained and their optimization problems are discussed analytically. The fatigue life of product merged on the best selection of material for better product development, considering economic aspect of present situation. In this book, the writer dives deeper into optimization technique to find out product life.

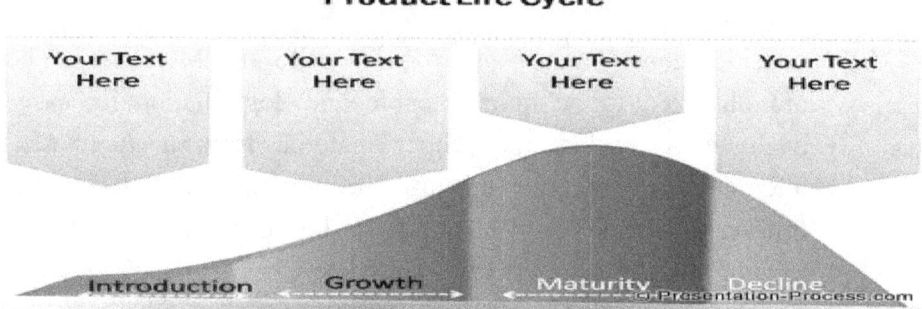

OVERVIEW OF GEAR MATERIAL:

Gears are commonly made of cast iron, steel, bronze, phenolic resins, acetal, nylon or other plastics. The selection of material depends on the type of loading and speed of operation, wear life, reliability and application. Cast iron is the least expensive. ASTM / AGMA grade 20 is widely used. Grades 30, 40, 50, 60 are progressively stronger and more expensive. CI gears have greater surface fatigue strength than bending fatigue strength. Better damping properties enable them to run quietly than steel.

Nodular cast iron gears have higher bending strength together with good surface durability. These gears are now a days used in automobile cam shafts. A good combination is often a steel pinion mated against cast iron gear. Steel finds many applications since it combines both high strength and low cost. Plain carbon and alloy steel usage is quite common.

Through hardened plain carbon steel with 0.35 - 0.6% C are used when gears need hardness more than 250 to 350 Bhn. These gears need grinding to overcome heat treatment distortion. When compactness, high impact strength and durability are needed as in automotive and mobile applications, alloy steels are used. These gears are surface or case-hardened by flame hardening, induction hardening, nitriding or case carburizing processes. Steels such as En 353, En36, En24, 17CrNiMo6 widely used for gears.

Bronzes are used when corrosion resistance, low friction and wear under high sliding velocity is needed as in worm-gear applications. AGMA recommends Tin bronzes containing small % of Ni, Pb or Zn. The hardness may range from 70 to 85Bhn. Non metallic gears made of phenolic resin, acetal, nylon and other plastics are used for light load lubrication free quiet operation at reasonable cost. Mating gear in many such applications is made with steel. In order to accommodate high thermal expansion, plastic gears must have higher backlash and undergo stringent prototype testing.

1.3 GEAR MATERIAL SELECTION MODELS:

Optimal design of gears requires the consideration of the two type parameters: Material and geometrical parameters. The choice of stronger material parameters may allow the choice of finer geometrical parameters and vice versa. Very important difference among these two parameters is that the geometrical parameters are often varied independently. On the other hand, material parameters can be inherently correlated to each other and may not be varied independently. An example of which being the variation of the bending fatigue limit (S_{bf}) with the core hardness (HB) for some steel materials. If these parameters would be varied independently in an optimization case, it may result in infeasible solutions.

Therefore, the final choice of material may not be possible within available data base. If gear material and geometrical parameters are optimized simultaneously then it is common to assume empirical formulas approximating a relation between material parameters for example the bending fatigue limit (Sbf) and ultimate tensile strength (UTS) as a function of hardness. If the choice of material is limited to a list of pre-defined candidates, then two difficulties can be appeared. First, a discrete optimization process should be followed against material parameters. Second, properties of different alternatives materials may not indicate any obvious correlation in the given list. The main goal is to choose material with best characteristic among alternatives.

Table 1. Shows suggested nine materials with their characteristics in a gear material selection.

RESEARCH AGENDA:

In an industry, design is a field that generally deals with different practices of design parameters; the research and development of processes, machine and equipment. The materials and process selection are key issues in optimal design of industrial products. Substituting and selecting materials for different machining parts is relatively common and often. Material selection is a difficult and subtle task, due to the immense number of different available materials. From this point of view paper deal with a set of major gear design criteria which are used for gear material selection. The main gear design criteria are: surface fatigue limit index, bending fatigue limit index, Surface Hardness, Core Hardness, Ultimate tensile strength, Cost, supply Lead Time. Using computer allows a large amount of information to be treated rapidly. One the most suitable models, for ranking alternatives gear materials, are SAW, MOORA, TOPSIS which using a multiple criteria, which all material performance indices and their uncertainties are accounted for simultaneously.

Industrial engineering is a branch of engineering dealing with the optimization of complex processes or systems. It is concerned with the development, improvement, implementation and evaluation of integrated systems of people, money, knowledge, information, equipment, energy, materials, analysis and synthesis, as well as the mathematical, physical and social sciences together with the principles and methods of engineering design to specify, forecast, and evaluate the results to be obtained from such systems or processes.

This paper concerns about increment the decision of material selection of gear manufacturing process and improvement the machinability, accuracy, quality, optimize the cost and time with the industrial view. Overall improvement of optimal design of a gear in manufacturing process considering the fatigue life and other aspect of materials.

1.5 OVERVIEW OF MCDM:

Multiple criteria decision making (MCDM) is the process of selecting the best alternative from a set of feasible alternatives considering multiple conflicting criteria. In precise terms criteria are considered to be 'strictly' conflicting if the increase in satisfaction of one results in a decrease in satisfaction of the other. An MCDM process always contains at least two alternatives and two conflicting criteria (Bhattacharya et al., 2003). MCDM are divided two broad categories: Multiple Attribute Decision Making (MADM) and Multiple Objective Decision Making (MODM). Several useful tools for solving of MCDM problems are,

- Simple Additive Weighting method (SAW)
- Technique for Order Preference by Similarity to Ideal Solution (TOPSIS)
- Multi Objective Optimization Ratio Analysis(MOORA)

- Analytical Hierarchy Method (AHP)
- Analytical Network Method (ANP) etc.

1.5.2 TECHNIQUE FOR ORDER PREFERENCE BY SIMILARITY TO IDEAL SOLUTION (TOPSIS)

TOPSIS is an evaluation method that is often used to solve MCDM problems [2, 3]. It has a number of applications [4, 5] in practice, such as comparison of company performances, financial ratio performance within a specific industry and financial investment in advanced manufacturing systems, etc. However, there are also some limits to it. So far, the work on how to improve original TOPSIS method has mainly emphasized on improving the weight to sensitize the R value [6, 7]. Besides, there has also been improvement on formula of the R value, such as the 'Miqiezhi' method [8]. Because of the complexity of evaluation problems, a better and simpler method is required to understand the inherent relationship between the R value and alternative evaluation. In this report, a novel, modified TOPSIS (M-TOPSIS) method is described as a process of calculating the distance between the alternatives and the reference points in the D+ D−-plane and constructing the R value to evaluate quality of alternative.

1.5.4 ENTROPY

Entropy was originally a thermodynamic concept, first introduced into information theory by Shannon (see Shannon, 1948 [21]). It has been widely used in the engineering, socioeconomic and other fields. According to the basic principles of information theory, information is a measure of system's ordered degree, and the entropy is a measure of system's disorder degree.

1.6 0VERVIEW OF MATLAB

MATLAB is a high-performance language for technical computing. It integrates computation, visualization, and programming in an easy-to-use environment where problems and solutions are expressed in familiar mathematical notation. Typical uses include:

• Math and computation

• Algorithm development

• Modeling, simulation, and prototyping

• Data analysis, exploration, and visualization

• Scientific and engineering graphics

• Application development, including graphical user interface building

MATLAB is an interactive system whose basic data element is an array that does not require dimensioning. This allows you to solve many technical computing problems, especially those with matrix and vector formulations, in a fraction of the time it would take to write a program in a scalar non interactive language such as C or Fortran.

The name MATLAB stands for matrix laboratory. MATLAB was originally written to provide easy access to matrix software developed by the LINPACK and EISPACK projects. Today, MATLAB uses software developed by the LAPACK and ARPACK projects, which together represent the state-of-the-art in software for matrix computation.

MATLAB features a family of application-specific solutions called toolboxes. Very important to most users of MATLAB, toolboxes allow you to learn and apply specialized technology. Toolboxes are comprehensive collections of MATLAB functions (M-files) that extend the MATLAB environment to solve particular classes of problems. Areas in which toolboxes are available include signal processing, control systems, neural networks, fuzzy logic, wavelets, simulation, and many others.

OBJECTIVE OF WORK

3.1 The proposed research work is planned into 5 stages:

    3.1.1 Identification of problem and setting up objective.

    3.1.2 Analysis of parameter and design of optimization tool.

    3.1.3 Effective simulation using MATLAB.

    3.1.4 Find fatigue life by construction of S-N diagram

    3.1.5 With the help of optimization tool rank the best alternatives (materials).

4.1.1 Phase 1:

Different objectives would be chosen form literature review for analysis and improvement, such as various alternative of gear materials and their criteria. Formation of MATRIX of gear materials to improve machinability, accuracy, quality, optimize the cost and time with the industrial view of a gear for better manufacturing product.

4.1.2 Phase 2:

Design is considered with proper selection of tool. For this research work MCDM is preferred as an optimization tool. But since the various method of MCDM is heavily used in material selection problem solving, hence might be some possible effective method can be added in this paper. In this stage more objective and harder matrix will be taken together.

A model of the proposed technique is presented below as flow diagram. This technique is a multiple criteria based decision making optimization technique which is mainly based on ranking to solve the problem and indicate the best selection of gear material.

4.1.3 Phase 3:
In this paper the problem solved by the MATLAB and showing the graph of materials and also detect the accuracy of the following problem.

4.1.3 Phase 4:
Find fatigue life by construction of S-N diagram

4.1.4 Phase 5:
With the help of optimization tool (SAW, MOORA & TOPSIS) rank the best alternatives (materials), by MATLAB software.

4. Selection of Gear Materials Considering Technical Economic and Supply Aspect by Ranking in MATLAB

4.1 PROBLEM DEFINITION
An organization has got 03 different materials with different specifications for gear. The decision maker considered 7 selection criteria. The materials are as follows:

MATERIAL	Grade	Surface (Bhn) Hardness (C1)	Core (Bhn) Hardness (C2)	Surface Fatigue Limit (MPa) (C3)	Bending Fatigue Limit (MPa) (C4)	UTS (MPa) (C5)	Cost (INR) Per kg (C6)	Supply Lead Time (In week) (C7)
Surface Hardened Alloy Steel (M6)	SAE 8620	542	229	1160	680	1850	60	6
Carburised steel (M7)	SAE 8620	647	297	1500	920	2300	60	5
Nitrided steel (M8)	EN40B	693	297	1250	760	1250	72	5

Table:01

5. Determination of Fatigue life of Three Best Gear Materials Considering Technical Aspect:

5.1 CALCULATING THE FATIGUE LIFE OF THREE BEST MATERIALS:

- Gear teeth act as a cantilever beam and it undergoes a fluctuating load so the bending fatigue limit is considerable.
- Manufacturing of gears undergoes a heat treatment process which is case hardening so surface fatigue limit of those materials are not to be considerable.
- For ferrous materials like steel, S-N curve becomes asymptotic at 10^6 cycles, which indicates the stress amplitude corresponding to infinite number of stress cycles.

The magnitude of this stress amplitude at 10^6 cycles represents the endurance limit of the materials.
- Considerable load of this experiment is 1000 MPa.

❖ Carburised Steel (M7)

CONSTRUCTION OF S-N DIAGRAM

- ✓ $σ_{ut}$ (MPA) = 2300
- ✓ $0.9\, σ_{ut}$ = 2070
- ✓ $Log10(0.9\, σ_{ut})$ = 3.31
- ✓ $σ_e$ = 920
- ✓ $Log10(σ_e)$ = 2.96
- ✓ Load = 1000 MPa

CALCULATION NUMBER OF CYCLE

$$EF = \frac{DB \times AE}{AD} = \frac{(6-3) \times (3.31-3)}{(3.31-2.96)}$$

$$= \frac{0.93}{0.35} = 2.66$$

$Log_{10}N = 3 + EF$

$= 3 + 2.66 = 5.66$

N = 457008 CYCLE

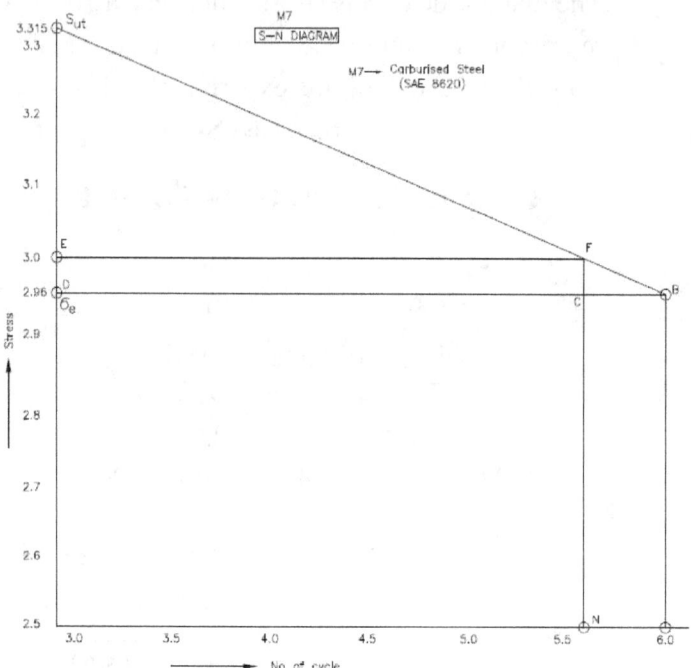

Fig:01

❖ **Nitrided Steel (M8)**

CONSTRUCTION OF S-N DIAGRAM

- ✓ σ_{ut} (MPA) = 1250
- ✓ $0.9\ \sigma_{ut}$ = 1125
- ✓ $\text{Log}10(0.9\ \sigma_{ut})$ = 3.05
- ✓ σ_e = 760
- ✓ $\text{Log}10(\sigma_e)$ = 2.88
- ✓ LOAD = 1000 Mpa

CALCULATION NUMBER OF CYCLE

$$EF = \frac{DB \times AE}{AD} = \frac{(6-3) \times (3.05-3)}{(3.05-2.88)}$$

$$=\frac{0.15}{0.17}=0.88$$

$Log_{10}N = 3+EF$

$=3+0.88=3.88$

N=7585 CYCLE

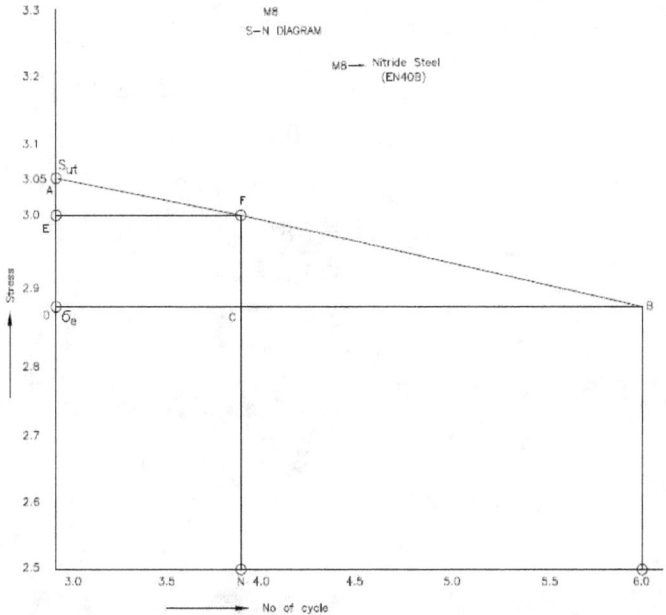

Fig:02

❖ **Surface Hardened Alloy Steel (M6)**

CONSTRUCTION OF S-N DIAGRAM

- ✓ σ$_{ut}$ (MPA)   =1850
- ✓ 0.9 σ$_{ut}$ =1665
- ✓ Log10(0.9 σ$_{ut}$)=3.22
- ✓ σ$_e$ =680
- ✓ Log10(σ$_e$)   =2.83
- ✓ LOAD   =1000 MPa

CALCULATION NUMBER OF CYCLE

$$EF = \frac{DB \times AE}{AD} = \frac{(6-3) \times (3.22-3)}{(3.22-2.83)}$$

$$= \frac{0.66}{0.39} = 1.69$$

Log$_{10}$N = 3 + EF

= 3 + 1.69 = 4.69

N = 489775 CYCLE

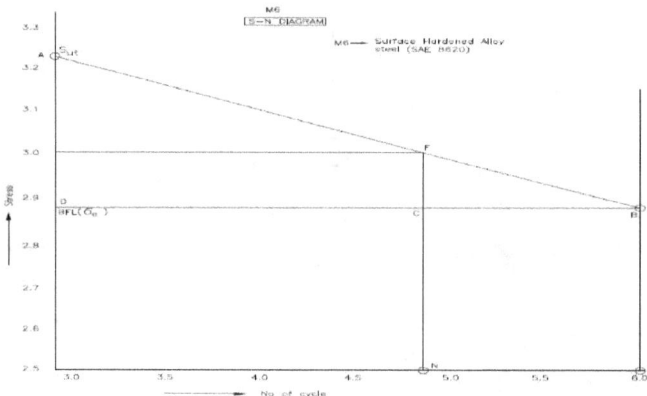

Fig:03

5.2 COMPARATIVE ANALYSIS ON NUMBER OF CYCLE OF GEAR MATERIALS USING S-N DIAGRAM:

MATERIAL	RANK	NO. OF CYCLE
Carburised Steel (M7)	1	457008 CYCLE
Nitrided Steel (M8)	2	7585 CYCLE
Surface Hardened Alloy Steel (M6)	3	489775 CYCLE

Table-4

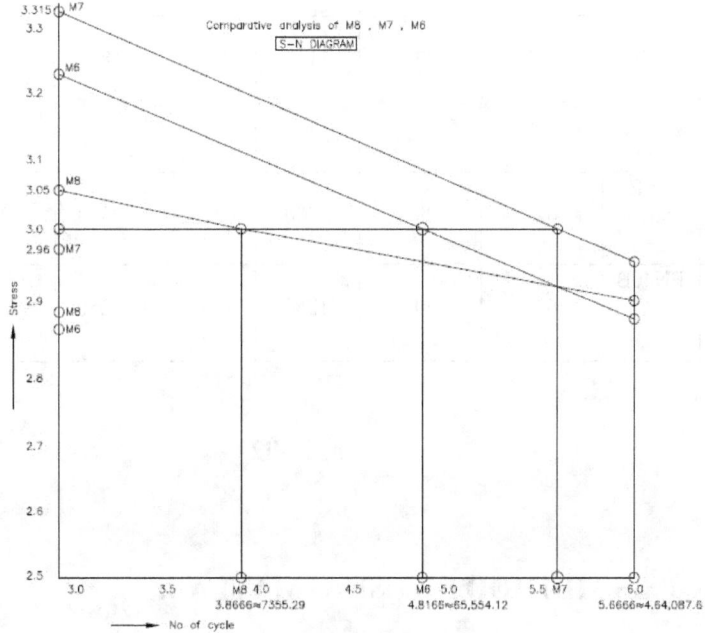

Fig:04

6. SELECTION OF BEST MATERIAL

6.1 FORMATION OF DEVLOPED MATRIX:

MATERIAL	Grade	Surface (Bhn) Hardness (C1)	Core (Bhn) Hardness (C2)	Surface Fatigue Limit (MPa) (C3)	Bending Fatigue Limit (MPa) (C4)	UTS (MPa) (C5)	Cost (INR) Per kg (C6)	Supply Lead Time (In week) (C7)	Number of cycles (C8)
Surface Hardened Alloy Steel (M6)	SAE 8620	542	229	1160	680	1850	60	6	489775 CYCLE
Carburised steel (M7)	SAE 8620	647	297	1500	920	2300	60	5	457008 CYCLE
Nitrided steel (M8)	EN40B	693	297	1250	760	1250	72	5	7585 CYCLE

Table-02

TOPSIS METHOD BY USING MATLAB:

RESULT:

The weighted values are:

0.1059   0.1080   0.1089   0.1100   0.1149   0.1022   0.1001   0.2501

The weighted values got from entropy method

STEP1: Determination of normalized decision matrix

0.7821   0.7710   0.7733   0.7391   0.8043   0.8333   0.8333   0.1122

0.9336   1.0000   1.0000   1.0000   1.0000   0.8333   1.0000   0.0158

1.0000   1.0000   0.8333   0.8261   0.5435   1.0000   1.0000   1.0000

STEP 2:

Determination of positive ideal solution: taking the maximum values of each column from the normalized decision matrix

1   1   1   1   1   1   1   1

Determination of negetive ideal solution: taking the minimum values of each column from the normalized decision matrix

0.7821   0.7710   0.7733   0.7391   0.5435   0.8333   0.8333   0.0158

STEP 3:

Calculation of the separation measure from the positive ideal solution(di_Plus)

0.4805

0.4955

0.1740

Calculation of the separation measure from the negetive ideal solution(di_Minus)

0.1007

0.2188

0.5096

STEP 3: Calculation of R_i

0.1732   0.3064   0.7454

Arranging the final value in descending order:--------->>>

M8 > M7 > M6

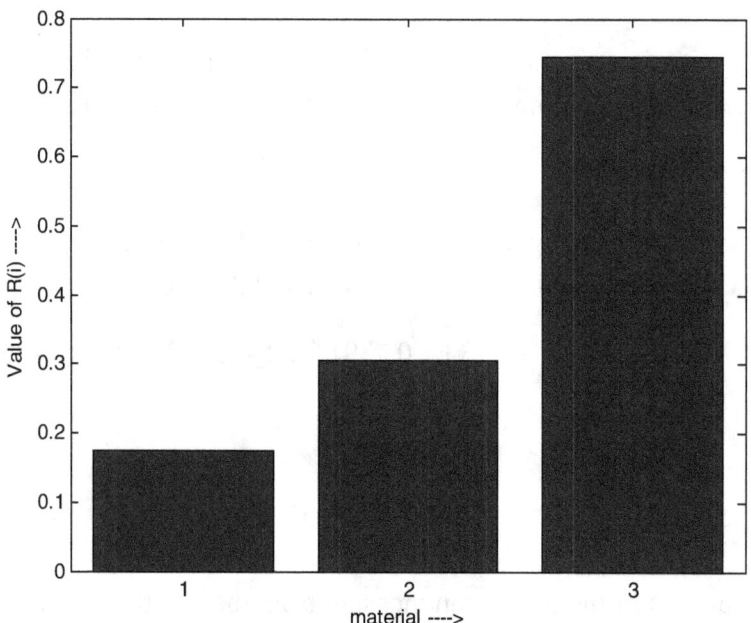

Fig: 05

7. CONCLUSION

It is quite clear that selection of a proper Gear Materials for a given manufacturing application involves a large number of considerations. The use of TOPSIS method is observed to be quite capable and computationally easy to evaluate and select the proper material from a given set of alternatives. These methods use the measures of the considered criteria with their relative importance in order to arrive at the final ranking of the alternative Gear Materials. Thus, these popular MCDM methods can be successfully employed for solving any type of decision-making problems having any number of criteria and alternatives in the manufacturing domain. Use of MATLAB software makes MCDM problem simple and gives prompt results which is very essential in today's decision making environment.

As far as design is concern fatigue life is very much important factor that influence the overall working life of the machine as well as the performance efficiency throughout its life span.

CHAPTER-3

Concept of applied Management in IE

3.1 Introduction of Total Quality Management (TQM):

Quality can be a compelling value in its own right. "It is robust enough to pertain to products, innovations, service standards, and caliber of people.... Everyone at every level can do something about it and feel the satisfaction of having made a difference. Making products that work, or providing first class service is something we can identify with from our own experience." (Pascale, 1991)

> A wide variety of approaches to defining quality are evident. For example:

i) Quality is defined as being about value (Feigenbaum, 1983)
ii) Quality is conformance to standards, specifications or requirements (Crosby, 1979)
iii) Quality is fitness for use (Juran, 1989)
iv) Quality as excellence (Peters and Waterman, 1982)
v) Quality is concerned with meeting or exceeding customer expectations (Parasuraman
   et al., 1985)
vi) Quality means delighting the customer (Peters, 1989)

> **Quality during the Industrial Revolution**

The industrial revolution revolutionized the manufacturing of products. Mass production set in large factories employing armies of people gave rise to new management ways. There were workers, supervisors and foremen, and managers. The establishment of factories and of this new organizational structure led to the withering of many small business trades and the removal of apprentices and masters from positions. Frederick Taylor's scientific management brought in efficient operations to increase output through mass production by breaking down jobs into parts with each part carried out by
Individual specialized workers. Practical use of Taylor's "scientific management," built around specialization and the division of labor, reached a high point with the advent of the mass production line with the workers performing repetitious tasks on a mammoth scale.

> **Mass Production and Scientific Management**

Mass production techniques reaped impressive early dividends. Henry Ford (1863-1947) built on the increased productivity brought by mass production. There was, however, more to Ford than flow lines and workers doing mindlessly repetitive tasks. Instead of controlling costs, to produce lower prices, Ford set the price and challenged the organization to ensure costs were low enough to meet the figure. The trouble was that when other manufacturers add extras, Ford lost touch with the aspirations of customers.

Compare this with the approach of Ford's predecessors. The first carmakers, such as France'sPanhard et Levassor, employed a small number of skilled craftsmen. The cars they produced were unique-- almost prototypes-- with parts being filed and cut to make them fit. As the parts were of varying sizes, craftsmanship was required. Ford bought in uniform and interchangeable parts. Skill departed and instead production was based round strict functional divides, i.e. demarcations. At the center of Ford's thinking was the aim of standardization-- something continually emphasized by the carmakers of today

though they talk in terms of quality and Ford in quantity. Scientific management emphasized the divorce of conception from execution and the substitutability of labor. The craftsmen concept disappeared with Taylorism and so did quality achieved through skilled craftsmanship. Inspection, thus, remained the sole guarantor of quality. Quality was no longer built into the product.

[source;]

- TQ / QM or TQM) and Six Sigma (6σ) are sweeping "culture change" efforts to position a company for greater customer satisfaction, profitability and competitiveness.
- We often think of features when we think of the quality of a product or service; TQ is about conformance quality, not features.
- Total Quality means—
  - ❖ Meeting Our Customer's Requirements
  - ❖ Doing Things Right the First Time; Freedom from Failure (Defects)
  - ❖ Consistency (Reduction in Variation)
  - ❖ Continuous Improvement
  - ❖ Quality in Everything We Do

3.2 DIMENSIONS OF QUALITY

The dimensions of quality primarily for manufactured products which a consumer looks for in a
Product includes the following:

Performance
The basic operating characteristic of a product is performance. For example, how well a car
Handles or its gas mileage etc.

Features
Features are the "extra" items added to the basic features, such as stereo CD or leather
interior in a car.

Reliability
Reliability is the probability that a product will operate properly within an expected time
frame, e.g., a TV without repair for about 7 years.

Conformance
Conformance is the degree to which a product meets pre-established standards.

Durability
Durability tells how long a product lasts, i.e. its life span before replacement.

Serviceability
Serviceability is the ease of getting repairs, the speed of repairs, and the courtesy and
competence of the repairperson.

Aesthetics
Aesthetics tells how a product looks, feels, sounds, smells or tastes.

Safety
Safety refers to assurance that the customer will not suffer injury or harm from the product. It
is an especially important consideration for automobiles.etc

3.3 Six sigma

Six Sigma is a process that enables companies to increase profits dramatically by streamlining operations, improving quality, and eliminating defects or mistakes in everything a company does, from raw materials to finish goods. A Six Sigma process generates a defect probability of 3.4 parts per million (PPM).

- Key activities in Six Sigma are:

1. Understanding customer needs (in quantifiable terms)
2. Translating the needs into the measurable outcomes

- Key objectives in Six Sigma are:
    1. Understanding & measuring the process inputs
    2. Looking at the root causes of variation

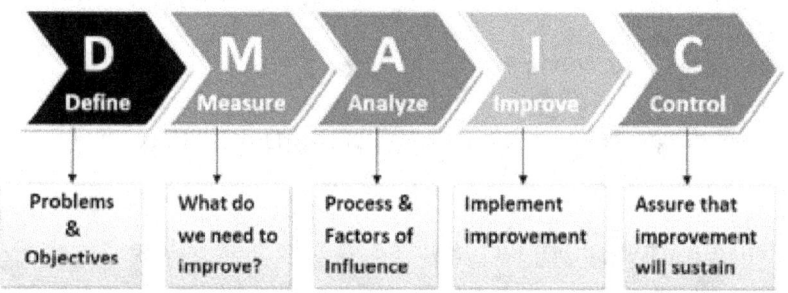

Timeline of Six Sigma Management

1978   Poor Quality!!! Motorola sells it TV business. When asked why, VP states "QUALITY STINKS!"

1980   Corporate Quality Officer appointed

1981   Training Center established

1985   Began to measure total defects/unit

1987   Corporation adopts Six Sigma program, Six Sigma goal to be achieved by 1992

1988   Motorola wins Malcolm Baldridge Award at the corporate level

1990   Six Sigma Research Institute formed in Ill.

Benefits of Six Sigma Management

- Improve process flows
- Reduce total defects
- Reduce process cycle time
- Enhance Customer and Employee satisfaction
- Help reduce inventory
- Help improve capacity and output
- Help increase quality and reliability
- Help decrease product costs
- Help improve product delivery to customer

Roles and responsibilities in Six Sigma Management

- Senior Executive
    - Provides the impetus, direction & alignment necessary for Six Sigma ultimate success. Senior Executive should:
        1. Study Six Sigma management
        2. Link company's objectives to Six Sigma projects
        3. Champion Six Sigma projects
        4. Constantly review Six Sigma projects progress
        5. Executive Committee Member
    - They are the top management of an organization. Executive Committee Members should:
        1. Deploy Six Sigma throughout the organization
        2. Prioritize and manage Six Sigma portfolio

3. Assign champion, BB and GB to Six Sigma projects
4. Remove barriers to Six Sigma management
5. Provide resources for Six Sigma management

- Champion
  - Take a very active sponsorship and leadership role in conducting and implementing Six Sigma projects. Champions should:
    1. Identify the project on the organizational dashboard
    2. Provide an ongoing communication link between the project team and Executive committee
    3. Keep the team focused on the project by providing direction and guidance
    4. Assure that Six Sigma methods and tools are being used in the project

- Master Black Belt
  - Takes a leadership role as keeper of the Six Sigma process and advisor to executives or business unit managers. Master Black Belt should:
    1. Counsel senior executives and business unit managers on Six Sigma management
    2. Continually improve and innovate the organization's Six Sigma process
    3. Apply Six Sigma across across both operations and transactions-based process
    4. Mentor Green Belts and Black Belts

- Black Belt
  - Is a full time change agent and improvement leader. Black Belts should have the following characteristics:

1. Technical and managerial process improvement/innovation skills
2. Understand the psychology of individuals and teams
3. Not intimidated by upper management
4. Has a customer focus

- The responsibilities of a Black Belt include:
    1. Communicate with the champion and process owner about progress of the project
    2. Help team members design and analyze experiments
    3. Provide training in tools and team functions to project team members
    4. Coach Green belts leading projects limited in scope

- Green Belt
    - Is an individual who works on projects part time, either as a team member for complex projects or as a project leader for simpler projects. Green Belts have the following responsibilities:
        1. Define & review project objective with project's champion
        2. Facilitate the team through all phases of the project
        3. Analyze data through all phases of the project
        4. Train team members in the use of Six Sigma tools and methods through all phases of the project

- Process Owner
    - Is the manager of a process. The process owner should be identified and involved in all Six Sigma projects relating to the process owner area. A process owner has the following responsibilities:

1. Empower employees to follow and improve best practice methods

2. Accept and manage the improved process after completion of the Six Sigma project

3. Understand how the process works, the capability of the process, and the relationship of the process to other processes in the organization

3.4 ISO-9000 Definition

"Quality the totality of features and characteristics of a product and service that bears on its ability to meet stated or implied needs."

- ☐ Quality is customer's determination. It is based on the customer's actual experience with the product or service, measured against his or her requirement.
- ☐ Quality consists of freedom from deficiencies.
- ☐ Quality is prevention constructing solutions to problems before they occur and designing excellence into a product or service.
- ☐ Quality is customer satisfaction the delight of the ultimate judge of how well products and services measure up.
- ☐ Quality is productivity from employees who receive the training, tools and instruction they need to execute their jobs.
- ☐ Quality is flexibility and willingness to change to meet demands.
- ☐ Quality is efficiency of doing things quickly and correctly.
- ☐ Quality is meeting a schedule and being on time.
- ☐ Quality is a process of ongoing improvement.
- ☐ Quality is an investment reaping a payoff because, in the long run, doing it right the first time is less expensive than correcting it later.
- ☐ Quality (knol'e-te')-- A systematic approach to the search for excellence. (Synonyms:productivity, cost reduction, schedule performance, customer satisfaction, teamwork and the bottom line).
- ☐ Quality always represents a moving target in a competitive market.

3.5 Industrial Engineers work to make things better, be they processes, products or systems

- **Typical focus areas include:**

Project Management
- Develop the detailed work breakdown structure of complex activities and form them into an integrated plan
- Provide time based schedules and resource allocations for complex plans or implementations
- Use project management techniques to perform Industrial Engineering analyses and investigations
- Conduct facility planning and facility layout development of new and revised production plants and office buildings
- Form and direct both small and large teams that work towards a defined objective, scope & deliverables
- Perform risk analysis of various project options and outcomes

Supply Chain Management
- Manage Supplier relationships
- Managing and report on company Supplier Cost/ Performance Indices to management
- Audit Suppliers and ensure supplier processes and procedures are being followed
- Travel to suppliers to resolve issues
- Coordinate first article Inspections
- Work with Outsource Manufacturers to ensure product quality, delivery and cost, is maintained

Program Management
- Develop proposals for new programs
- Manage program/project teams to ensure program stays on schedule, on budget, and meets performance expectations
- Coordinate a matrix of team member across departments within an organization to ensure completion of project tasks

Ergonomics/Human Factors
- Ensure Human Factors Engineering is utilized in New Product Design

- Ensure Human Factors Engineering disciplines are utilized in production setup and configuration
- Ensure company Ergonomics policies are defined to minimize causes of employee injury and discomfort

Technology Development and Transfer
- Identify basic business problems requiring analysis
- Determine if technology or process based solution best
- Characterize problem, identify prospective providers/bidders and submit requests for proposals
- Evaluate bid responses, select successful bidder(s) and establish technical feasibility
- Conduct small scale/medium scale tests to determine operational feasibility, implementation methods and training requirements
- Conduct enterprise wide implementation
- Transition support activities/responsibilities to long term business and technology owners

Management of Change
- Ensure that change programs are coordinated, support one another and move along the critical path
- Create and maintain the imperative for the change, establish priorities and provide visible
Sponsorship for the change
- Provide the skills, knowledge, processes, organization structure and tools required to deliver the change
- Ensure that the individuals involved buy into the change, actively support it and adopt their behavior accordingly

Financial Engineering
- Determine production costs using specific cost based methodology
- Develop budgets, forecasts for operating cost centers
- Measure actual performance vs budget goals and investigate variance
- Develop capital and expense budgets for capacity expansion
- Perform cost analysis/justification for capital and expense expenditures
- Perform make vs buy vs lease analyses

Strategic Planning
- Develop long range planning models, typically 5-10 years in scope
- Model all areas affected by operation

- Identify anticipated investment in plant, capacity, network, etc
- Tie to preliminary production cost, operational cost, sales forecasts
- Develop preliminary financial impacts, including profitability and ROI

3.8 Concept of Inventory Management

Inventory management is a significant component of any business since inventories are normally accountable for the majority of the costs incurred in business operations. Most companies meet the customer demand by tracking and maintaining the inventory required using an inventory management system. Traditionally, Controlling Inventory requires a typical problem to be balanced between diverse departments. The most significant industry segments in terms of opportunities in the field of 'Inventory Control', production and retail. Managing inventory is a factor of utmost importance for the overall functioning of the organizations because it involves a huge capital investment and can result in painful consequences due to both over-stocking as well as under-stocking. In the following report we have shown 3 inventory models namely Simple EOQ Model, EOQ Model with stock out allowed and Inventory Model under Risk based on the raw materials used in the mechanical workshop of Haldia Institute of Technology. EOQ is Economic Order Quantity with basically permits lowest cost per unit and Inventory models determines when and how much inventory to carry. This project goes through the process of analyzing the workshop's current Forecasting model and recommending an inventory control model. As a result, an Economic Order Quantity (EOQ) and a Reorder Point was recommended along with forecasting techniques to help them reduce their product stock outs.

INTRODUCTION

Inventory is the detailed list of those movable items which are necessary to manufacture a product and to maintain the equipment and machinery in good working order. The quality and the value of every item is also mentioned in the list. Inventory is any stored resource that is used to satisfy a current or future need. Raw materials, work-in-process, and finished goods are examples of inventory. Two basic questions in inventory management are (1) how much to order (or produce), and (2) when to order (or produce). With today's uncertain economy, companies are searching for alternative methods to keep ahead of their competitors by effectively driving sales and by cost reduction. Big retail companies do not stand a chance in today's environment if they do not have an appropriate inventory control model intact. The Economic Order Quantity and a Reorder Point (EOQ/ROP) model have been used for many years, but yet some companies have not taken advantage of it. An Economic order quantity could assist in deciding what would be the best optimal order quantity at the company's lowest price. Similar to EOQ, the reorder point will advise when to place an order for specific products based on their historical demand. The reorder point also allows sufficient stock at hand to satisfy demand while the next order arrives due to the lead time. Since retail can be unpredictable and competitive, it will be interesting to see how forecasting can affect the economic order quantity (EOQ) and reorder point. EOQ is Economic Order Quantity with basically permits lowest cost per unit and Inventory models determines when and how much inventory to carry. This project goes through the process of analyzing the workshop's current Forecasting model and recommending an inventory control model.

INVENTORY MODELS

Concept:

> ➢ Inventory models determine when and how much inventory to carry.

- Inventory models handle chiefly two decisions:
    1) How much to order at one time, and
    2) When to order this quantity to minimize total costs.
- Lowest cost decision rules for inventory management pertain to either buying products from outside or producing them within the company.
- *Simple inventory models* assume no delivery delay and that demand is known.
- *Probabilistic models* handle situations of risk and uncertainty.

Concept: A problem which always remains is that how much material may be ordered at a time. An industry making bolts will definitely like to know the length of steel bars to be purchased at any one time. This length of steel bars is called "ECONOMIC ORDER QUANTITY" and an economic order quantity is one which permits lowest cost per unit and is most advantageous.

Before calculating economic order quantity it is necessary to become familiar with terms like maximum inventory, minimum inventory, standard order and recorder point, which are known as Quantity Standards.

Starting from an instant when inventory OA is in the stores, it (inventory) consumes gradually in quantity from A along AD at a uniform rate. It is preknown that it takes L number of days between initiating order and receiving the required inventory. Therefore as the quantity reaches point B, purchase requisition is initiated which takes from B to C, that is time R. From C to D is the inventory procurement time P. At the point D when only reserve stock is left, the ordered material is supposed to reach and again the total quantity shoots to its maximum value, i.e. the point A`(A=A`).

Maximum Quantity OA is the upper or maximum limit to which the inventory can be kept in the stores at any time.

Minimum Quantity OE is the lower or minimum limit of the inventory which must be kept in the stores at any time.

The purpose should be to hold enough and not excessive stock of material. *Stock holding:*

a) Avoids running out of stock.
b) Helps creating a buffer stock which may be utilized if the material falls below the minimum level.
c) Makes sure the predecided delivery dates.
d) Provides quick availability of materials.
e) Takes care of price fluctuations and shortage of inventory in the market.
f) Advises regarding, obsolete and slow moving items.
g) Helps in standardization and thus reducing the variety of items to be handled.

Standard Order(A`D) is the difference between maximum and minimum quantity and it is known as economical purchase inventory size.

Reorder Point (B) indicates that it is high time to initiate a purchase order and if not done so the inventory may exhaust, and even reserve stock utilized before the new material arrives.

From B^1 to D^1 it is as *lead time (L)* and it may be calculated on the basis of past experience. It includes:

a) Time to prepare purchase requisition and placing the order;
b) Time taken to deliver purchase order to the seller;
c) Time for seller(vendor) to get or prepare inventory;
d) Time for the inventory to be dispatched from the vendor's end and to reach the customer.

Time (a) above is known as requisition time (R) and (b) +(c) + (d) is the procurement time (P).

The economic lot size for an order or the economic order quantity depends upon two types of costs:

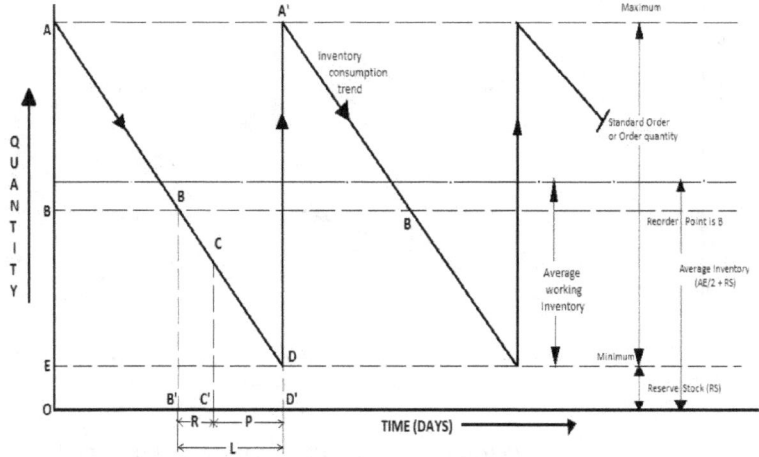

FIG 1: QUANTITY STANDARDS

(a) ***Inventory procurement cost*** which consists of expenditure connected with:

1. Receiving quotations
2. Processing purchase requisition
3. Following up and expediting purchase order
4. Receiving material and then inspecting it
5. Processing seller's invoice

Procurement costs decrease as the order quantity increases (*FIG. 2*)

(b) ***Carrying costs***, which vary with quantity ordered, based on the average inventory and consists of:

1. Interest on capital investment;
2. Cost of storage facility, up-keep of material record keeping etc.
3. Cost involving deterioration and obsolescence; and
4. Cost of insurance property tax etc.

Carrying costs are almost directly proportional to the order size or lot size or order quantity,

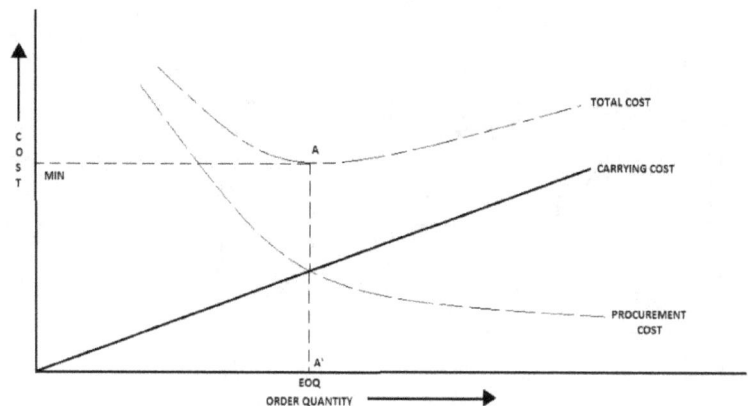

FIG. 2: RELATIONSHIP BETWEEN COST & QUANTITY

In *FIG.2* the procurement costs and inventory carrying costs have been plotted with respect to quantity in lot. Total cost is by adding procurement cost and carrying cost. Total cost is minimum at the point A and thus A` represent the economic order quantity or economic lot size.

Another method of finding E.O.Q that is by mathematical means is given below:

Let, **Q** is the economic lot size or E.O.Q.
C is the cost for one item.
I is the cost of carrying inventory in percentage per period, including insurance,
obsolescence, taxes etc
P is the procurement cost associated with one order.
and **U** is total quantity used per period say annually.

Number of purchase orders to be furnished

$$= \frac{Total\ Quantity}{EOQ} = \frac{U}{Q}$$

Total procurement cost = Number of purchase orders ×cost involved in one purchase or

procurement = $\dfrac{U}{Q} \times P$

Average annual inventory = Q/ 2

Inventory carrying cost = Average inventory × cost per item × cost of carrying inventory in percent per period.

$$= \dfrac{Q}{2} \times C \times I$$

Total cost, T = (a)+(b)

$$T = [U \times P/ Q] + [Q \times C \times I /2]$$

$$T = U.P.Q^{-1} + \dfrac{Q.C.I}{2}$$

To minimize that total cost, differentiate T w.r.t Q and put it equal to zero

$$\dfrac{dT}{dQ} = \dfrac{d}{dQ}\left(U.P.Q^{-1} + \dfrac{Q}{2} \times C \times I\right)$$

$$0 = -U.P.Q^{-2} + \dfrac{C.I}{2}$$

Or, $\dfrac{U \cdot P}{Q^2} = \dfrac{C.I}{2}$

Or, $Q^2 = \dfrac{2.U.P}{C.I}$

Or, $Q = \sqrt{\dfrac{2.U.P}{C.I.}}$

ABC Analysis

ABC analysis is a type of analysis of material dividing in three groups called A-group items, B-Group items and C-group items For the purpose of exercising control over materials. Manufacturing concerns find it useful to divide materials into three categories.

An analysis of the annual consumption of materials of any organisation would indicate that a handful to top high value items (less than 10 per cent of the total number) will account for a substantial portion of about 70 per cent of total consumption value. Graphical Representation of ABC Analysis

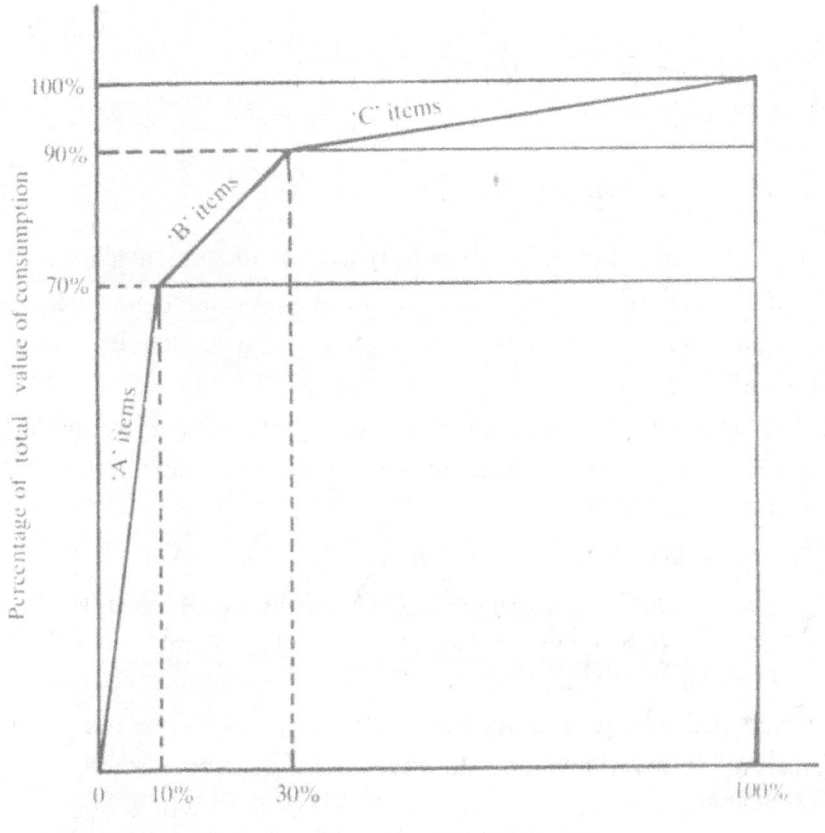

Graphical Representation of ABC Analysis

Remember: 10% of total number of items carries 70% of value. - "A" group items

Similarly, a large number bottom items (over 70 per cent of the total number of items) account for only about 10 percent of the consumption value.

Remember: 70% of total number of items account for only about 10% of consumption value - "C"-group items.

Between these two extremes will fall those items the percentage number of which is more or less equal to their consumption value.

Remember: 20% of total number of items account for only about 20% consumption value - "B" group items.

Items in the top category are treated as "A" items, items in the bottom category are called as "C" category items and the items that lie between the top and the bottom are called "B" category items. Such an analysis of materials is known as ABC analysis or Proportional parts value analysis.

Classification of items into A, B and C categories

The logic behind this kind of analysis is that the management should study each item of stock in terms of its usage, lead time, technical or other problems and its relative money value in the total investment in inventories.
Critical items i.e., high value items deserve very close attention and low value items need to be devoted minimum expense and effort in the task of controlling inventories.
The Material Manager by concentrating on "A" class items is able to control inventories and show visible results in a short span of time. By controlling "A" items and doing a proper inventory analysis, obsolete stocks are automatically pinpointed.
ABC analysis also helps in reducing the clerical costs and results in better planning and improved inventory turnover. ABC analysis has to be resorted to because equal attention to A, B and C items will not be worthwhile and would be very expensive.

The following steps will explain to you the classification of items into A, B and C categories.

The following steps will explain to you the classification of items into A, B and C categories.

1. Find out the unit cost and and the usage of each material over a given period.
2. Multiply the unit cost by the estimated annual usage to obtain the net value.
3. List out all the items and arrange them in the descending value. (Annual Value)
4. Accumulate value and add up number of items and calculate percentage on total inventory in value and in number.

5. Draw a curve of percentage items and percentage value.
6. Mark off from the curve the rational limits of A, B and C categories.

MATERIAL REQUIREMENTS PLANNING (MRP)

- MRP is a computational technique that converts the master schedule for end products into a detailed schedule for raw materials and components used in the end products. The detailed schedule identifies the quantities of each raw materials components items. It also tells when each item must be ordered and delivered so as to meet the master schedule for the final products.
- The purpose of MRP is to ensure that materials and components are available in the right quantities and at the right time so that finished products can be completed according to the master production schedule.
- MRP is often considered to be a subset of inventory control. It is an effective tool for minimizing unnecessary inventory investment.
- MRP is useful in production scheduling and purchasing of materials.
- The concept of MRP is relatively straight forward.

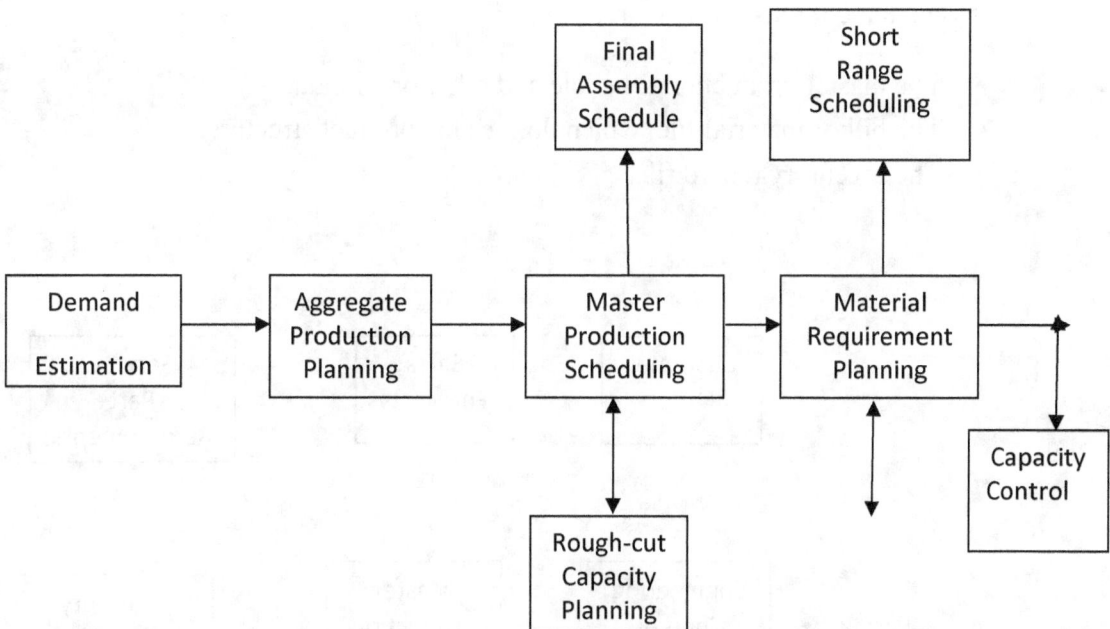

> Detailed capacity Requirement Planning

FUNCTIONS:

An MRP system has three major functions/uses:

- Control of inventory levels,
- Assignment of priorities for components, (depending upon their delivery dates) and
- Determination of capacity requirements at a detailed level.

Capacity Requirements planning(CRP) is a system for determining if a planned production schedule can be accomplished with available capacity and, if not, making adjustments as necessary.

INPUTS TO MRP

MRP converts the master production schedule into the detailed schedule for raw materials and components. For the MRP program to perform its function, it must operate on the data contained in the master schedule. However, this is only one of three sources of input data on which MRP relies. The three inputs to MRP are:-

- The master production schedule and other order data.
- The bill of material file, which defines the product structure.
- The inventory record file.

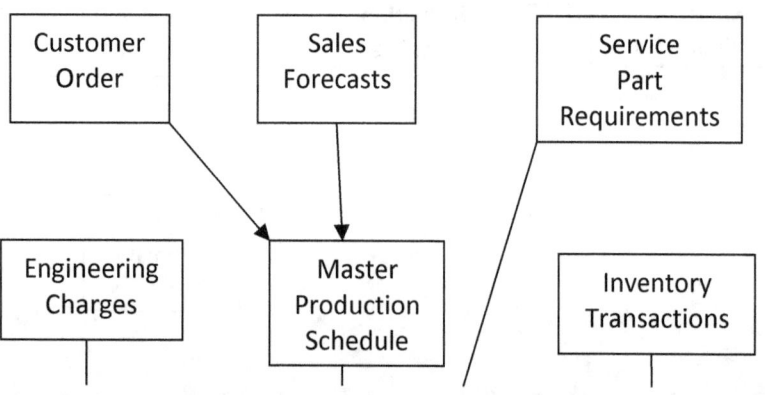

Master Production Schedule

- ♣ The master schedule is based on an accurate estimate of demand for the firm's product, together with a realistic assessment of its production capacity.
- ♣ The master production schedule is a list of what end product are to be produced, how many of each product is to be produced, and when the products are likely to be ready for shipment.
- ♣ The general format of a master production schedule is illustrated in below figure. It may show weekly or monthly delivery schedules.

Week number	5	6	7	8	9	10	11
Product P1				50		100	
Product P2				70	80	25	
etc							

Bill of materials file

- ♣ In order to compute the raw materials and components requirements for end products listed in the master schedule, the product structure must be known. This is specified by the bill of materials which is a list of components parts and sub assemblies that make up each product. Putting all these assembly lists together, we have the bill of materials file (BOM).
- ♣ Below figure shows the structure of an assembled product. It shows, subassembly S1 is the parent of components C1, C2 and C3. Products P1 is the parent of sub assemblies S1 and S2 and so on.

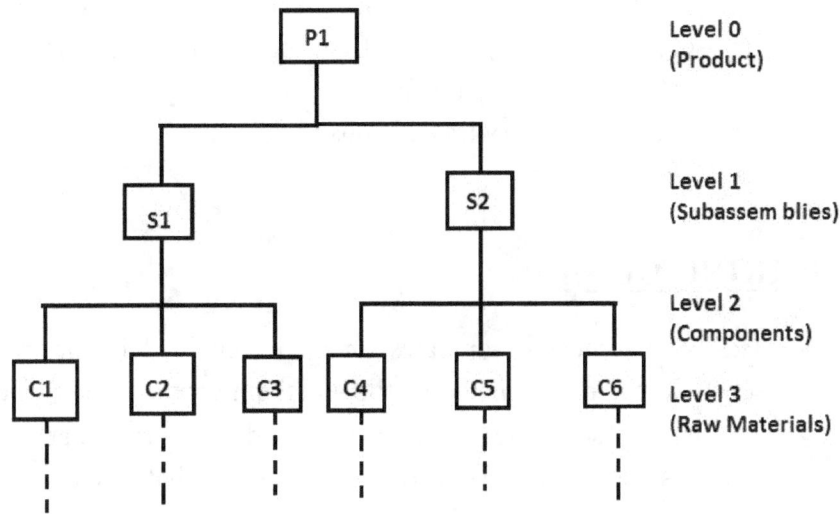

The product structure must also specify how many of each item is included in its parent. This is accomplished in above figure by the number in parentheses to the right and below each block. For example, subassembly S1 contains four of components C2 and one each of components C1 and C3.

INVENTORY RECORD FILE

- It is mandatory in MRP to have accurate current data on inventory status. This is accomplished by utilizing a computerized inventory system which maintains the inventory record file or item master file.
- A definition of the lead time for the raw material, components and assemblies must be established in the inventory record file. The ordering lead time can be determined from purchasing records. The manufacturing lead time can be determined from the process route sheets.

It is important that the inputs to the MRP processor be kept current. The bill of materials file must be maintained by feeding any engineering changes that affect the product structure into the BOM. Similarly, the inventory record file is maintained by inputting the inventory transactions to the file.

WORKING OF MRP

- The MRP processor operates on data contained in the masters' schedule, the bill of materials file and the inventory record file.
- The master schedule specifies a period by period list of final products required. The BOM defines what materials and components are needed for each product. The inventory record file contains information on the current and future inventory status of each component.
- The MRP program computer- how many of each component and raw materials are needed by exploding the end product requirement into successively lower levels in the product structure.
- There are several factors that must be considered in the MRP parts and materials explosion.

    1. Components and subassemblies already existing in the stock must be considered for determining requirements for meeting the master schedule.
    2. The MRP processor must determine when to start assembling the subassemblies by offsetting the due dates for these items by their respective manufacturing lead times. Similarly, the components due dates must be offset by their manufacturing lead times. The MRP program performs these lead time offset calculation from data obtained in the inventory record file and from route sheet data.
    3. Some components and many raw materials are common to several products. The MRP processor must collect these

common use items during the explosion. The total quantities for each common use item are then combined into a single net requirement for the item.
4. Since master production schedule provides time-phased delivery requirement for the end products, this time phasing must be carried through the calculation of the individual components and raw materials requirements.

MRP OUTPUT PEPORTS

The MRP program generates a variety of outputs that can be used in the planning and management of plant operations. These outputs includes:-

a. Primary Outputs:
   1. Order release notice, to place orders that have been planned by the MRP system.
   2. Reports showing planned orders to be released in future periods.
   3. Cancellation notices, indicating cancellation of open orders because of changes in the master schedule.
   4. Rescheduling notices, indicating changes in due dates for open orders.
   5. Reports on inventory status.

b. Secondary outputs
   1. Performances reports of various type, indicating costs, item usage, actual versus planned lead times etc.
   2. Exception reports, showing deviations from schedule, orders that are overdue, scrap and so on.
   3. Inventory forecasts, indicating projected inventory levels in future periods.

BENEFITS OF MRP

1. Reduction in inventory.

2. Improved customer services- late orders are reduced as much as 90%.
3. Quicker response to changes in demand.
4. Better machine utilization.
5. Greater productivity

RAW MATERIALS

Raw Materials used in the workshop that are kept in inventory are :

SL. NO.	ITEMS	QUANTITY	ITEMS SPECIFICATION	ORDERING TIME
1.	Welding Electrode	1000 units	Ø 3.15 × L 305	
2.	Welding Plates	800 -	75×50×5	
3.	Hacksaw Blade	100 -		
4.	Metallic Bar	500 -	Ø 40 × L 85	
5.	Bentonite	100 k.g		
6.	Calcium Oxide	100 k.g		
7.	Linseed Oil	2 L		
8.	Charcoal	150 k.g		
9.	Tool Holder	10 units		
10.	Cutting Tool	20 -		

DESCRIPTION OF ITEMS

WELDIND ELECTRODE: These are conductors, through which electricity enters into the welding materials in the form of heat with coatings that can consist of a number of different compounds, including rutile, calcium fluoride, and cellulose and iron powder. Rutile electrodes, coated with 25%–45% TiO_2, are characterized by ease of use and good appearance of the resulting weld. The composition of the electrode core is generally similar and sometimes identical to that of the base material. The electrode is coated in a metal mixture called flux, which gives off gases as it decomposes to prevent weld contamination, introduces deoxidizers to purify the weld, causes weld-protecting slag to form, improves the arc stability, and provides alloying elements to improve the weld quality. Electrodes can be divided into three groups—those designed to melt quickly are called "fast-fill" electrodes, those designed to solidify quickly are called "fast-freeze" electrodes, and intermediate electrodes go by the name "fill-freeze" or "fast-follow"

WELDING PLATES: Welding plates or the base metals are the metals or plates that are to be joined or cut. The weld metals that are solidified in the joint can be only of base metal or a mixture of base metal and filler metal.

HACKSAW BLADE: A hacksaw is a fine-tooth saw with a blade held under tension in a frame, used for cutting materials such as metal or plastics. Hand-held hacksaws consist of a

metal arch with a handle, usually a pistol grip, with pins for attaching a narrow disposable blade. A screw or other mechanism is used to put the thin blade under tension. The blade can be mounted with the teeth facing toward or away from the handle, resulting in cutting action on either the push or pull stroke. On the push stroke, the arch will flex slightly, decreasing the tension on the blade.

METALLIC BAR: Metallic bars are the mostly used material in any workshop. These bars are generally made of iron. Metallic bars are mostly used for the purpose of various lathe operations like turning, facing, taper turning etc.

BENTONITE: Bentonite is an absorbent aluminum phyllosilicate, essentially impure clay consisting mostly of montmorillonite. There are different types of bentonite, each named after the respective dominant element, such as potassium (K), sodium (Na), calcium (Ca), and aluminum (Al). For industrial purposes, two main classes of bentonite exist: sodium and calcium bentonites. Bentonite slurry walls are used in construction, where the slurry wall is a trench filled with a thick colloidal mixture of Bentonite and water.

CALCIUM OXIDE: Calcium oxide (CaO), commonly known as quicklime or burnt lime is a widely used chemical compound. It is a white, caustic, alkaline crystalline solid at room temperature. The broadly used term lime connotes calcium-containing inorganic materials, in which carbonates, oxides and hydroxides of calcium, silicon, magnesium, aluminum, and iron predominate, such as limestone. By contrast, quicklime specifically applies to a single chemical compound.

LINSEED OIL: Linseed oil, also known as flaxseed oil, is a clear to yellowish oil obtained from the dried ripe seeds of the flax plant. The oil is obtained by cold pressing, sometimes followed by solvent extraction. Linseed oil is a "drying oil", as it can polymerize into a solid form. Due to its polymer-forming properties, linseed oil is used on its own or blended with other oils, resins, and solvents as an impregnator and varnish in wood finishing, as a pigment binder in oil paints, as a plasticizer and hardener in putty and in the manufacture of linoleum. The use of linseed oil has declined over the past several decades with the increased use of synthetic alkyd resins, which function similarly but resist yellowing. It is edible oil but, because of its strong flavor and odor, is only a minor constituent of human nutrition. When used as a wood finish, linseed oil dries slowly and shrinks little upon hardening. Linseed oil does not cover the surface as varnish does, but soaks into the (visible and microscopic) pores, leaving a shiny but not glossy surface that shows off the grain of the wood. A linseed oil finish is easily repaired, but it provides no significant barrier against scratching. Only wax finishes are less protective. Liquid water will penetrate a linseed oil finish in mere minutes and water vapor bypasses it almost completely.

CHARCOAL: Charcoal is the dark grey residue consisting of carbon, and any remaining ash, obtained by removing water and other volatile constituents from animal and vegetation substances. Charcoal is usually produced by slow pyrolysis, the heating of wood or other substances in the absence of oxygen (see pyrolysis, char and biochar). It is usually an impure form of carbon as it contains ash. Charcoal burns at intense temperatures, up to 2700 degrees Celsius. By comparison the melting point of iron is approximately 1200 to 1550 degrees Celsius. Due to its porosity it is sensitive to the flow of air and the heat generated can be moderated by controlling the air flow to the fire. For this reason charcoal is an ideal fuel for a forge and is still widely used by blacksmiths. Historically, charcoal was used in great quantities for smelting iron in bloomeries and later blast furnaces and finery forges.

TOOL HOLDER: Tool holders are the device to hold the cutting tool in the correct position with respect to work piece, and provide enough holding force to counteract the cutting forces acting on the tool, example tool post.

CUTTING TOOL: The sizes of these tools are generally square or rectangular in cross section. The shank is that part of the tool on one end of which the cutting point is formed. It is supported in the tool post of the lathe. The base is that part of the shank which bears against the support and bears the tangential force of the cut.

Economic Order Quantity

Concept: A problem which always remains is that how much material may be ordered at a time. An industry making bolts will definitely like to know the length of steel bars to be purchased at any one time. This length of steel bars is called "ECONOMIC ORDER QUANTITY" and an economic order quantity is one which permits lowest cost per unit and is most advantageous.

Types of EOQ models:

  I. Simple E.O.Q. model
  II. E.O.Q model with stockouts allowed,

III. Inventory models under risk.

(I) Simple EOQ model:

- The simple EOQ model can be used if the demand is known with certainty.
- The demand and lead time are known.
- The item will be purchased from outside and that demand will continue well into the future.
- It is also assumed that not only the demand is known with certainty, but that is the same from day to day and that stock outsare not allowed.
- If we start to observe the inventory position immediately after receipt of an order, the quantity in stock Q decreases steadily until a lead time's supply is reached. If lead time (L) is 5days and demand is 4 (pieces) per day, a lead time's supply is 20 pieces or units. Therefore when 20 units remain in the store,an order is placed.This is called the re-order point.Exactly 5 days after the order is placed,the stock is replenished and the cycle repeats itself.

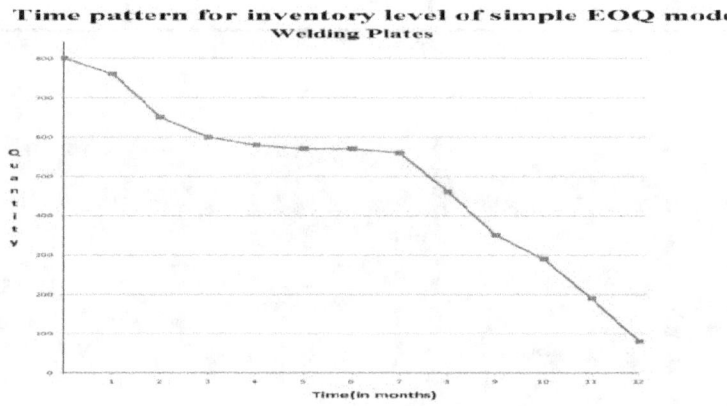

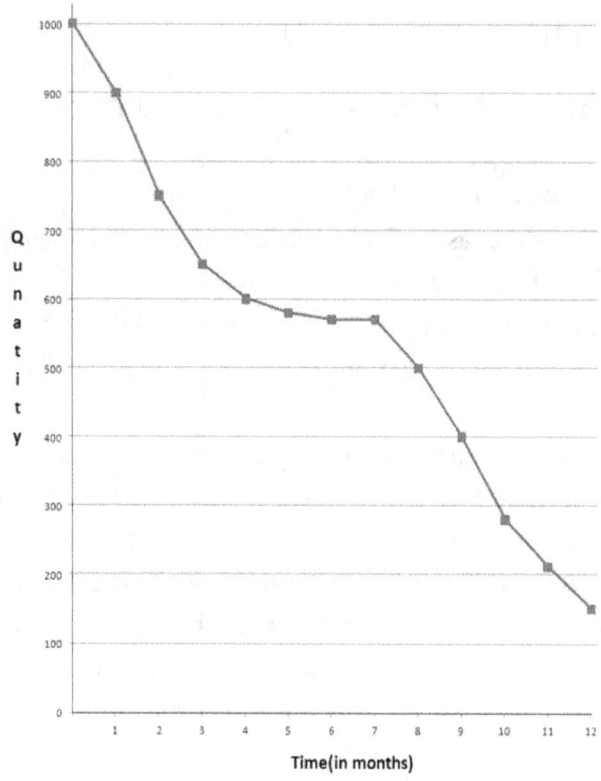

Time pattern for inventory level of simple EOQ model for Welding Electrode

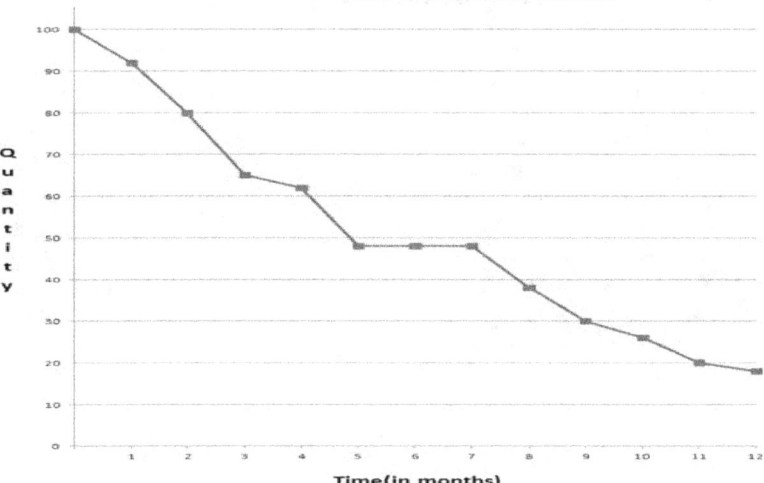

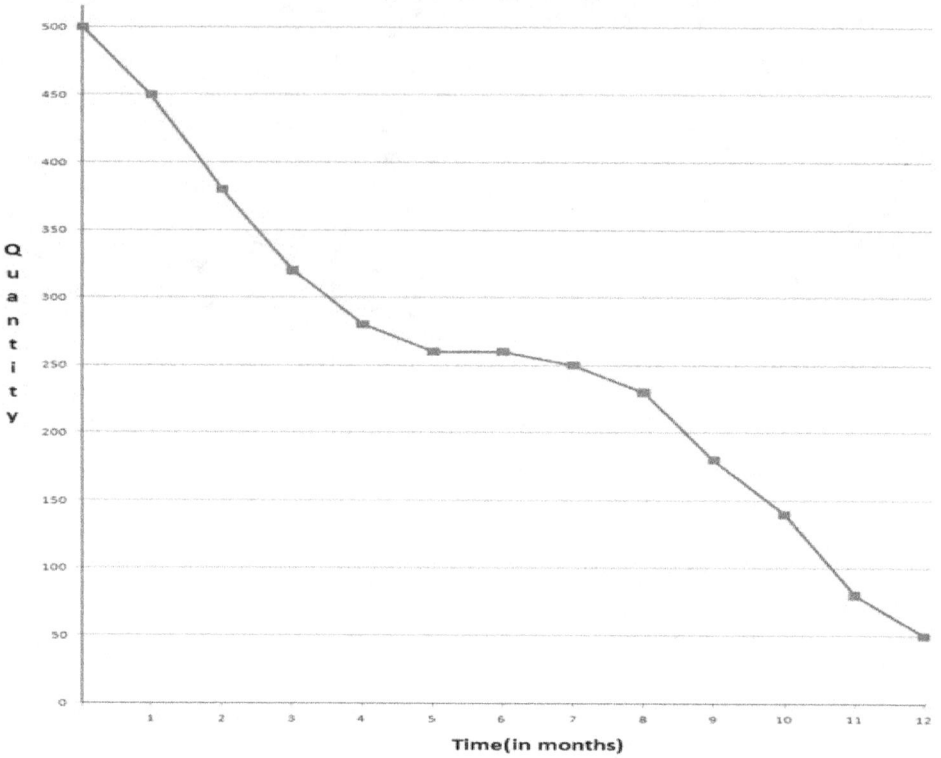

(II) EOQ model with stockout allowed:

- **Stockout** means running out of stock.
- In simple EOQ model, the stock was always available.
- The out of stock position will lead to a back order. This situation is frequently found in mailing-order houses where an item is temporarily out-of-stock and the customer is willing to wait until it is replenished. The firm does suffer some cost in this situation, since, at the very least, some expediting and extra communication with the customer, must take place. In addition to the explicit costs, repeated backorders will certainly lead to an erosion of goodwill.

In some situations, backorder may be economically justified.Forinstance,with very high value items such as commercial jet planes,no inventory is carried and a backorder state always exists.

- The second possible outcome of an out-of-stock position is the lost sales.Here the cost is much more severe.In this case, the customer places an order,receives an out-of-stock response,and take his business elsewhere.The company must take into account the likelihood that the customer will not return and that therefore the profit on future sales might also be lost.In addition,the loss in goodwill that is will precipitate may also persuade others to purchase elsewhere.
- Since backorder and lost sales costs are hard to estimate, service levels are frequently specified.For example, management may feel that stockouts should not occurmare than 2% of the time.

In this case, $$Q = \sqrt{\frac{2UP}{C.I}}\sqrt{\frac{CI+B}{B}} \quad \text{and} \quad M = \frac{QB}{CI+B}$$

Where,**Q** is EOQ

U is Annual use

P is Procurement cost per order

C is cost per piece

I is cost of carrying inventory

B is incurred for every backorder

M is maximum inventory.

Hacksaw Blade EOQ Model with stockout allowed

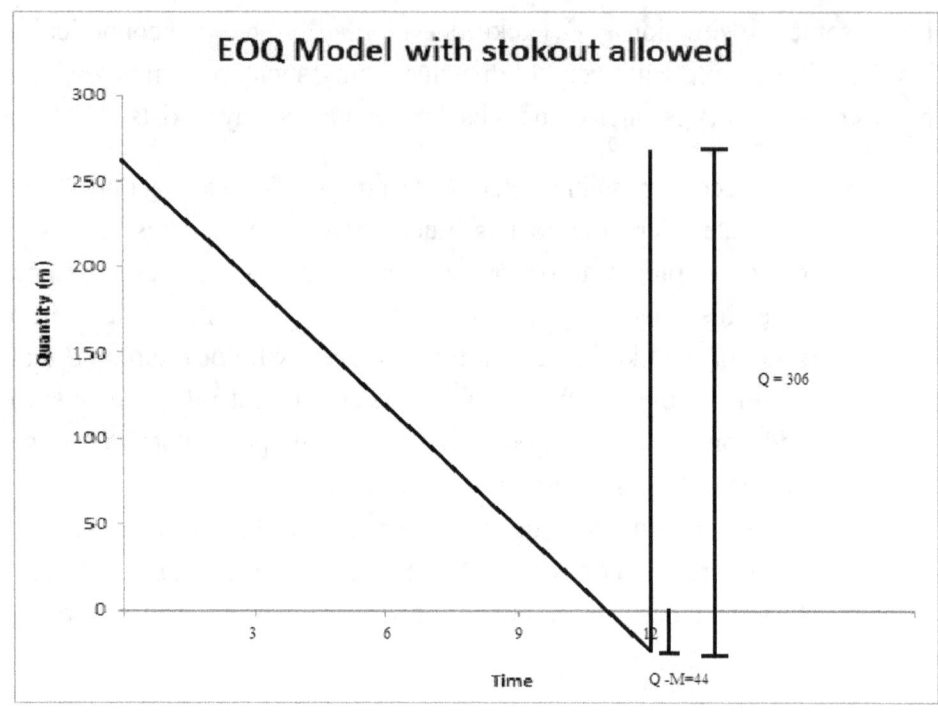

U is Annual use= 100 units

P is Procurement cost per order =Rs. 2000

C is cost per price =Rs.25

I is cost of carrying inventory, a percentage – including insurance,obsolescence, taxes etc. =20%

B is cost incurred for every backorder = Rs.30

M is maximum inventory =?

Q= √ (2UP/CI) × √ {(CI+B)/B}

=√ (2×100×2000/5) × √ {(5+30)/30}

=306

M = (QB/CI+B)

= (306×30/5+30)

=262

So stockout is (Q – M)= 306 –262 =44

Welding Electrode EOQ Model with stokout allowed

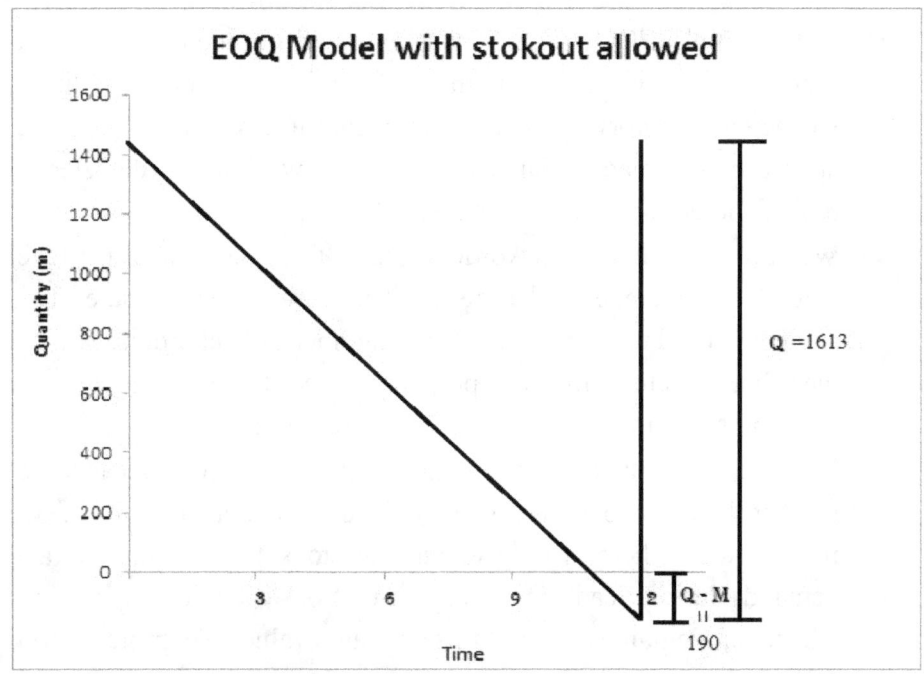

U is Annual use = 1000

P is Procurement cost per order = Rs. 2000

C is cost per price = Rs.10

I is cost of carrying inventory, a percentage – including insurance,obsolescence, taxes etc.

= 20%

B is cost incurred for every backorder = Rs.15

M is maximum inventory =?

Q = √ (2UP/CI) × √((CI+B)/B)

=√ (2×1000×2000/2) × √ ((2+15)/15)

=1613

M = (QB/CI+B)

=(1613×15/2+17)

=1423

So stockout is (Q – M) = 1613 – 1423 =190

(III)Inventory model under risk:

- In the simple EOQ model it was assumed that demand and lead time were both known with certainty. Under this condition, whenever inventory levels reached a lead time's worth of demand, an order was placed. Then as the stock was finally depleted, a replenishment order would arrive.
- We did find, that if backorders had a finite cost, a order level below that average lead time demand may be appropriate. This strategy would result in some desire number of backorders. We can therefore conclude that the purpose of reorder is not to prevent stock outs, but to keep them within desired limits.
- In the case of fluctuating demand, it may be quite logical to set reorder levels above the average lead time demand; for if we reorder when there is only enough in stock to meet the average demand during lead time, an out-of-stock position will occur whenever demand exceeds this average value. To protect from undesirably large stock out situations, safety stocks are maintained. They will provide the cushion needed whenever demand exceeds this average.
- Safety stocks (OS) can be defined as the difference between the reorder level and the average lead time demand. Therefore as the reorder point is raised, the safety stock increases and the likelihood of a stock out during any cycle decreases.
- An order is placed whenever stock levels fall to *"r"*. On the average, stock will be depleted to *"s"* when the replenishment order arrives. During the first cycle we see that demand was about average; however in the second cycle the demand was much greater than average. Protecting the system from a stock out during the lead time, was a safety stock maintained at *s*. If more protection was desired, the level of "r" could have been increased. With a higher *"r"*, the safety stock would have been larger and the chances of a stock out occurring in any one cycle would have been lower.

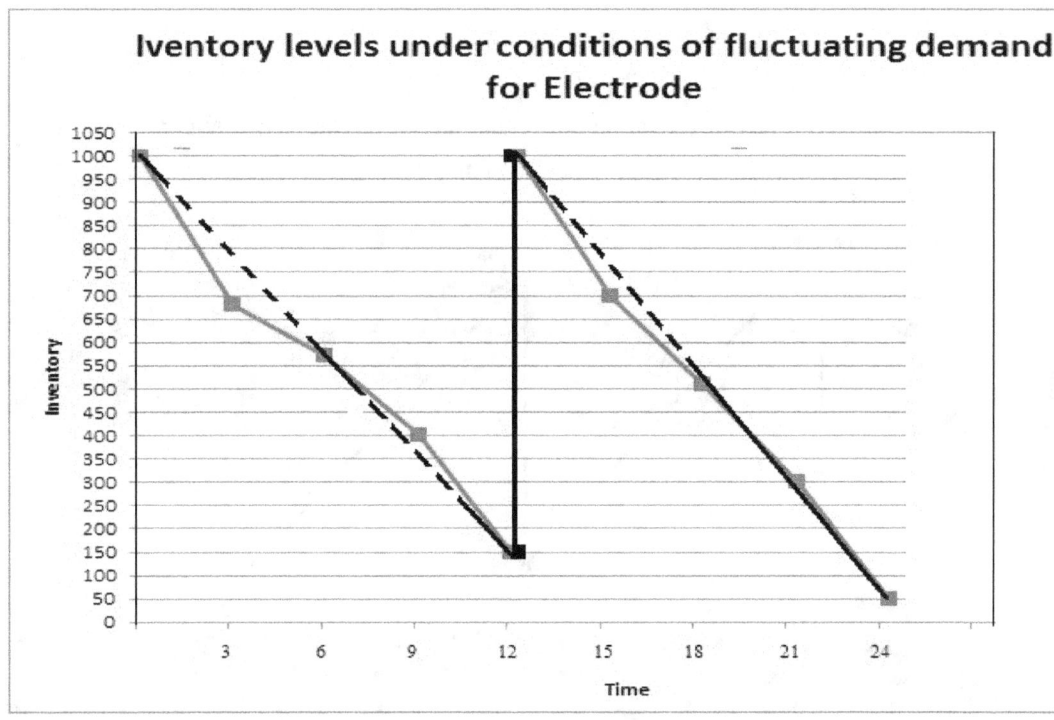

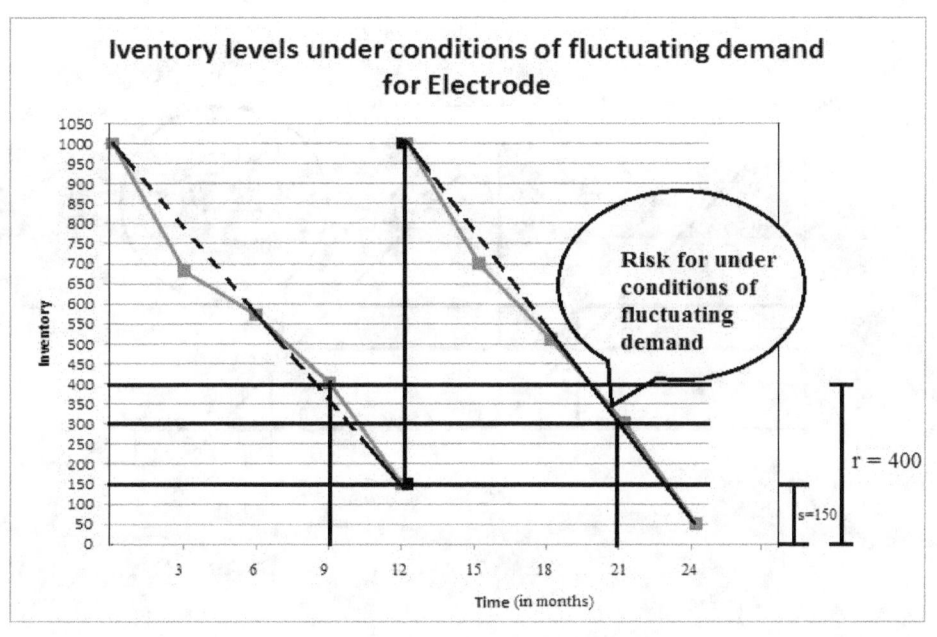

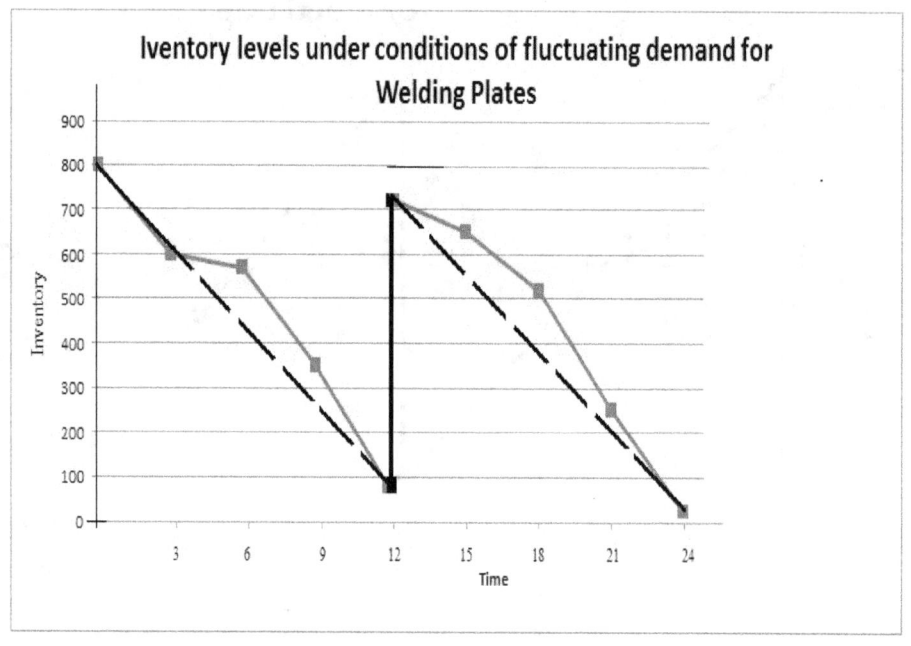

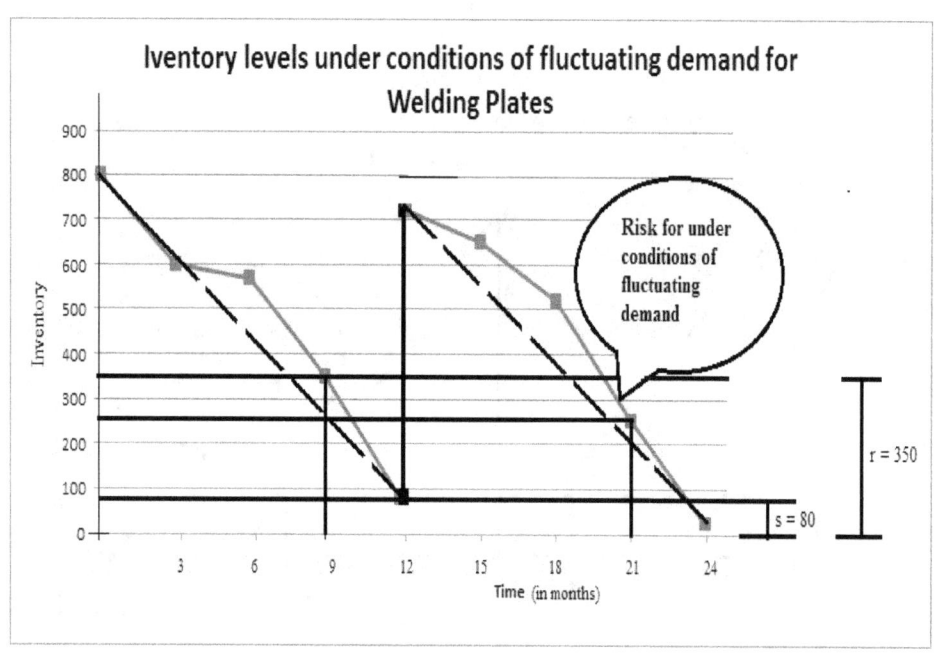

CHAPTER-4

Industrial Engineering in NDT environment

4.1 Concept of Inspection:

An inspection is, most generally, an organized examination or formal evaluation exercise. In engineering activities inspection involves the measurements, tests, and gauges applied to certain characteristics in regard to an object or activity. The results are usually compared to specified requirements and standards for determining whether the item or activity is in line with these targets, often with a Standard Inspection Procedure in place to ensure consistent checking. Inspections are usually non-destructive.

- Objective is to achieve economic and un-interrupted plant operation by ensuring availability of safe, reliable and statutory complied static equipments.

> **Inspection in petrochemical/oil refinery industries:**

The oil and gas industry operates according to stringent standards that primarily strive to keep equipment running efficiently while maintaining workplace safety. Conforming to such standards is challenging because oil and gas operations involve drilling (both on land and offshore), reservoir engineering, well servicing, production services, refining and transportation of petroleum products, and many other operations.

Not only is it challenging to conform to industry standards, but it is expensive to maintain a safe workplace and to protect workers from harm. It is necessary to equip workers with protective apparel; to continually train them on practicing safety procedures; and to keep them up to date on practicing improved equipment inspection techniques.

4.2 Role Industrial Engineers in Inspection:

- The responsibility of Inspection Engineers in a refinery is very critical. Inspection Engineers can contribute in a great way to improve profitability. The major contributing areas include
    - Routine Inspection and timely recommendation
    - Periodic shutdown inspection and maintenance
    - RLA study and health forecast.
    - Metallurgy Upgradation
    - To inspect, measure and record the deterioration of material and to evaluate present physical condition of the equipments and its components for their To recommend / forecast short term and long term repairs and replacements to ensure further run-length.
    - To initiate procurement action of material to meet recommended repair / replacement needs.
    - To maintain upto date maintenance and inspection records and history of equipments.
    - soundness to continue in service.

> **Plant Availability factor (PAF)**

- Plant Availability Factor is calculated as per the formula=No. of days the unit is available for operations / 365 days. This means if a unit is one operation for X days & shutdown for planned & emergency maint for Y days & idle for Z days, then The PAF = (365-Y)/365

Onstream Factor

- Onstream Factor = (365-Y-Z) / 365

4.3 Plant Shutdown:

A **'plant shutdown'** means the closure of a factory. **Plant** is a synonym for a factory or an industrial building or a production unit. So a **plant shutdown** means closing down or shutting down of a **plant** or a factory.

> **Shutdowns are normally of two types**

I. Maintenance & Inspection Shutdowns (Planned & emergency)
II. Decoking shutdowns

I. Maintenance & Inspection Shutdown

- To minimise the duration of shutdown, Inspection should review history of all critical equipment
- Recommend well in advance all major repair anticipated in the shutdown
- Arranging all critical material, I.e. follow up.
- Job list should contain minimum number of equipment to be taken up in shutdown.
- Inspection of all repair jobs to ensure quality

II. Decoking shutdown

- The high temp units like Coker, Visbreaker where cracking of heavier hydrocarbons take place face coking
- Like in Coker, Coking phenomena is called delayed coking. This means the coking is not allowed in the heater tubes but it is shifted to the Reactor.
- Even then some coking cannot be ruled out in the heater tubes or transfer line.
- So, to operate the unit efficiently decoking is done for these units.
- Decoking interval is between 6 to 12 months depending on its operating conditions.
-

4.4 Five Most Popular Inspection Methods:

An Overview:

The purpose of this article is to outline 5 popular methods: visual inspection, ultrasonic techniques, radiography, thermography and acoustic emissions.

Each of these methods is explained, followed by a qualitative discussion of its implementation.

Method 1: Visual Inspection

Visual inspection is an inexpensive method for detecting equipment flaws and defects. A trained technician is likely to detect improper structural installations, certain types of impending structural failure, welding flaws, corrosion development, and cracks.

Method 2: Ultrasonic techniques

Ultrasonic testing utilizes sound waves whose frequencies (50 kHz - 50 MHz) are above the audible range for the human ear.

The piezo-electric effect of the ultrasonic transducer makes it possible to transmit and receive from within the equipment. The instrument makes it **possible to inspect the internal** structure of the equipment, and to detect thickness changes, welds, cracks, voids, delamination and other types of material or structural defects.

The limitation of this method is that data acquisition and evaluation depends on the expertise of the technician. This makes it <u>difficult</u> to arrive at nonsubjective readings and precision.

Method 3: Radiography

Radiographic methods utilize X-ray or gamma rays (electromagnetic radiation) to examine the internal structure and integrity of the equipment. Because these waves have short wave lengths, they can penetrate and travel through structural materials such as steel and metallic alloys.

In the oil and gas industry, this NDT method is useful for inspecting welds on pipelines and pressure vessels. It is also useful for inspecting non-metallic materials such as concrete and ceramics. Operating this type of NDT requires conformance to safety regulations.

Method 4: Thermography

Thermographic inspection measures the difference between the temperature of a pipeline and the surrounding environment. The measurement helps to detect (a) defects in pipeline insulation, and (b) leakage of oil or gas.

The marriage of this NDT technique and drone services will make this method of inspection more efficient and cost effective.

Method 5: Acoustic emissions

This method detects the presence of rarefaction waves produced by leaks in pipelines. When a fluid leak occurs, negative pressure waves propagate in both directions within the pipeline. Detection of these acoustic waves helps identify leakage in pipelines.

[source: http://info.industrialskyworks.com/blog]

4.5 What mean is NDT?

Many NDT (Non Destructive Testing) methods are utilized in the oil and gas industry. The best NDT methods address issues regarding safety, equipment reliability, and environmental protection and government regulations. The greatest benefits that NDT service provides are that

(a) Equipment for transporting petroleum products (such as a pipeline) can be inspected without making any structural changes;

(b) Equipment is not disturbed during NDT; therefore there is neither a reason to shut down nor to interrupt operations.

Popular NDT methods involve visual inspections, ultrasonic techniques, radiography, thermography, laser shearography, eddy current testing, microwaves, and acoustics.

4.6 Nondestructive Evaluation:

Nondestructive evolution comprises nondestructive testing, nondestructive inspection and nondestructive examination.

it includes testing, inspection & examination to determine some characteristic of the object or to determine the condition of the object.

It is used to find, locate size or determine the flaws in the object & to decide whether the flaws are acceptable or not. A flaw that is rejected is called a defect.

> ### Selection of Nondestructive Evaluation:

Nondestructive evaluation can be divided into following areas,

- Flaw detection & evaluation
- Leak detection & evaluation
- Dimensional measurement & evaluation
- Structure / microstructure evaluation
- Estimation of mechanical & physical properties
- Chemical composition determination
- Stress/ strain analysis

4.7 Brief Description of Applied NDT Methods:

I. Visual Inspection

Most basic and common inspection method. Tools include fiberscopes, borescopes, magnifying glasses and mirrors. Robotic crawlers permit observation in hazardous or tight areas, such as air ducts, reactors, pipelines. Portable video inspection unit with zoom allows inspection of large tanks and vessels, railroad tank cars, sewer lines.

Robotic crawlers permit observation in hazardous or tight areas, such as air ducts, reactors, pipelines.

Function:

- Done when object is accessible,

- Mirors / magnifying lens may be used,
- Proper illumination is a vital requirement,
- Inspector shall have annual visual exmination,
- Remote visual inspection using telescopes / boroscopes / fibre optics / cameras. Such instruments should have resolution capability equivalent to direct visual inspection.

II. Dye Penetrant Test (DPT):

- A liquid with high surface wetting characteristics is applied to the surface of the part and allowed time to seep into surface breaking defects.

- The excess liquid is removed from the surface of the part.

- A developer (powder) is applied to pull the trapped penetrant out the defect and spread it on the surface where it can be seen.

- Visual inspection is the final step in the process. The penetrant used is often loaded with a fluorescent dye and the inspection is done under UV light to increase test sensitivity.

> ### STEPS FOR DPT

Surface preparation: application of claener,
Application of penetrant,
Time allowed for penetration,
Removal of excess penetrant from the surface,
Application of developer,
Inspection & interpretation,

\# Final cleaning

If discontinuities are observed, same to be removed by grinding etc. And above steps to be repeated for the defective locations till become acceptable.

> ➢ **Penetration Time (Dwell Time)**

Penetration time is the period for which penetrant is permitted to remain on the specimen. This is a vital parameter for the test.

Minimum time required for the penetrant to enter into the discontinuity is determind by,

- \# Manufacturer's recommendation,
- \# Type of material tested,
- \# Type of discontinuity expected,
- \# Temperature of the specimen,
- \# Humidity of the environment

for gross discontinuity, it takes 3-5 minutes.

For tight crack type defects, it may be 30 min.

> ➢ **Removal of Excess Penetrant:**

After sufficient dwell time, excess penetrant shall be removed by lintfree cloth / rag.

In case of fluorescent penetrant, removal is seen by black light

Interpretation:

- \# Continuous / intermittent line indications are caused by cracks, cold shuts, forging laps etc.
- \# Round indications are caused by gas holes, holes, pin holes, porosity

➤ Advantages & Limitations of DPT

Advantages:

ANY SURFACE FLAWS CAN BE DETECTED REGARDLESS OF SIZE, CONFIGURATION, INTERNAL STRUCTURE / CHEMICAL COMPOSITION, FLAW ORIENTATIONETC. OF THE TEST PIECE.

CAN BE USED FOR MULTILAYER WELDING INSPECTION IN STAGES

Limitations:

ONLY IMPERFECTIONS / DISCONTINUTIES OPEN TO THE SURFCE CAN BE DETECTED,

ROUGH & POROUS SURFACE MAY PRODUCE FALSE INDICATIONS,

UNSUITABLE FOR LOW DENSITY POWDER METALLURGY PARTS OR OTHER POROUS MATERIALS

III. Magnetic Particle Testing

It is a non-destructive test method to detect surface & sub-surface discontinuties in ferro-magnetic materials.

➤ Principle:

Magnetic flux leak occurrson the surface over the defects when a magnetic material having discontunities is magnetised. Fine magnetic particlesapplied on the area get attracted & held in true form & orientation of the defects.

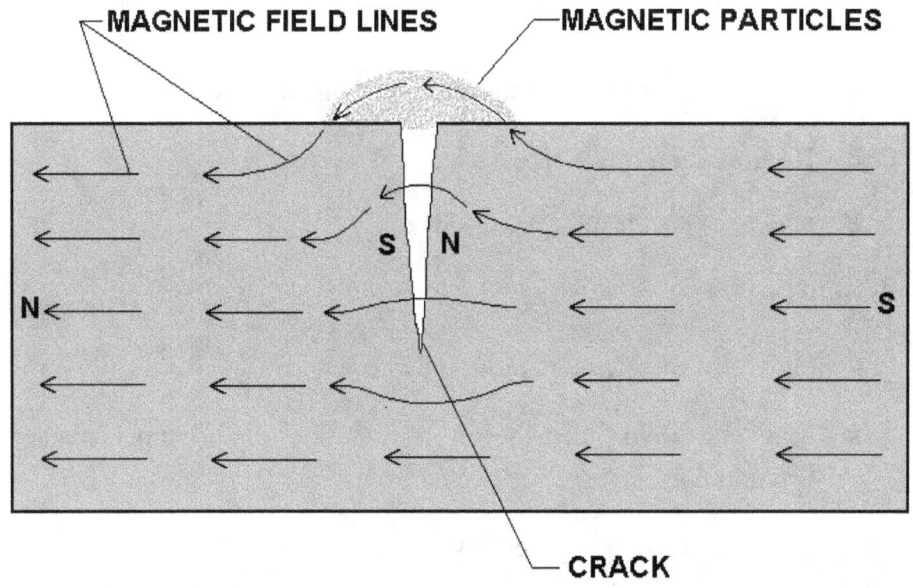

> **Magnetisation Method & Current**

Method:

By passing electric current through material

By placing the material in a magnetic field by an external source,

Part may be magetised fully / partially as per requirement.

Current:

Ac

Single phase half wave rectified ac (hwac)

Three phase full wave rectified ac (fwdc)

For detection of surface indication, ac shall be used. For detection of sub-surface indication hwdc & fwdc shall be used.

Advantages & Limitations

Advantages:

- Very sensitive for locating small & shallow indications
- No limitation of the size & shape of the part
- No elaborate cleaning required.

Limitations:

- Only ferromagnetic materials can be tested,
- Magnetic field must be in a direction of principal plane of discontunoty,
- Demagnetisation is necessary,
- Skill needed in interpretation
- Non-magnetic coating (paint/ plating) affect sensitivity of inspection

Type Of Magnetic Particles

Dry powder of different colours suiting to background like black, grey, red, brown etc. Which is applied in low velocity using powder blower or sprinkler,

In wet method, magnetic ink (black & brown) / fluorescent ink is applied by spraying.this method can not be used on hot objects.

Ink vicosity: 3cst, flash point: 57deg.c min.

Wet method:

- Excellent for detection of sub-surface defect,
- Good mobility when used with ac or hwdc

\# Gives clear indication

Dry method:

Not so sensitive as wet method for detection of surface & sub-surface defect, also slower.

Required property:-

- \# Non toxic,
- \# Finely divided, ferromagnetic,
- \# High permeability,
- \# Low retentivity,
- \# High colour contrast,
- \# Correct size & mobility

IV. **Ultrasonic Testing**

In this method, sound wave of very high frequency is employed. Its property is similar to light wave i.e. it can be reflected & refracted.

- ➢ Mechanical vibration (ultrasound) is introduced into a test piece, travels through the piece and gets reflected from the surface perpendicular to its travel direction. Reflected wave travels through the same path & speed but in opposite direction & enters the receiver.
- ➢ Frequency range for inspection: 1 to 25 mhz.

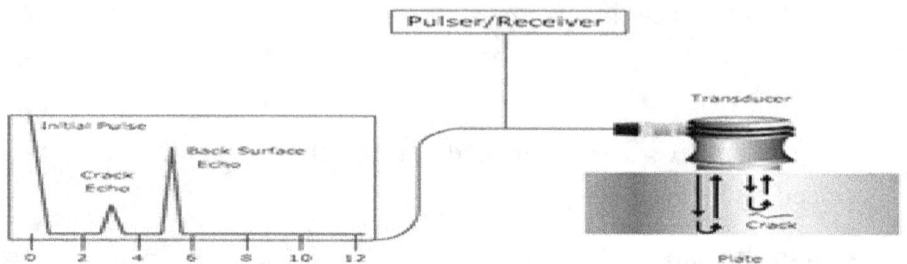

> **Uses of Ultrasonic Testing:**

It is the most widely used NDT method. Primary uses are,

\# To measure thickness,

\# To measure extent of corrosion,

\# Detection & chracterisation of internal flaws

\# To detect surface flaws.

> **Advantages:**

\# Upto 20' thk. Material may be examined,

\# Highly sensitive to detect very small flaw,

\# Greater accuracy than other ndt,

\# Only one surface need to be accessible,

\# Very fast,

\# Volumetric scanning ability,

\# Not hazardous,

\# Portability

> **Limitations of Ultrasonic Testing**

- \# Requires extensive technical knowledge,
- \# Irregular, very thin & non-homogenous substances are difficult to inspect,
- \# Defects just below the surface may not be detectable,

V. Radiography Testing

The radiation used in radiography testing is a higher energy (shorter wavelength) version of the electromagnetic waves that we see as visible light. The radiation can come from an X-ray generator or a radioactive source.

> **Principle:**

It is based on the differential absorption of penetrating radiation. Because of the differences in density, variations in thickness or differences in absorption characteristic due to variations in compositions, different amount of radiation is absorbed. Unabsorbed radiation passing through the part received on film / photosensitive paper or viewed on fluorescent screen / monitor.

> **Method:**

Three elements of radiography are,

- \# The test piece or object,
- \# Radiation source,
- \# Recording medium (film)

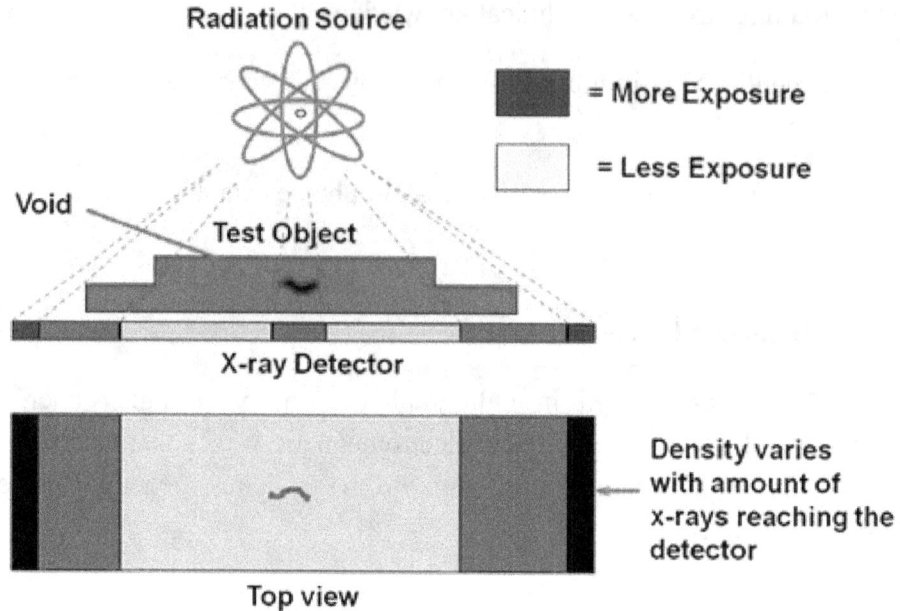

Absorption of the radiation passing through the specimen will be different by the flaws & the balance portion of the specimen.

Thus, amount of radiation reaching on the film beneath the flaw will be different from rest of the area.

This produces a latent image of the flaw on the film.

On developing the film, a **shadow** of defferent photographic density is seen due to the flaw than the surrounding area.

> **Variable Parameters:**

\# Radiation source

\# Type of film

Source to film distance (sfd)
Radiation beam alignment

VI. THERMOGRAPHY:

It is temperature based technique which produces thermal image of any object based on infrared radiation emitted by them.

The infrared region of the electromagnetic spectrum covers wavelength from 0.75 to 12 micron. Thermographic range is between 3 to 6 micron for short wave and 6 to 12 for long wave.

The infrared (ir) radiation emitted by an object under observation is received by a camera unit & transmitted by an ir optical system to an ir detection element. The detector converts the scanned ir radiation into a video signal. The signal is processed in the display unit where a live thermal picture called thermogram is presented.

VII. EDDY CURRENT TESTING

Eddy current testing is particularly well suited for detecting surface cracks but can also be used to make electrical conductivity and coating thickness measurements. Here a small surface probe is scanned over the part surface in an attempt to detect a crack.

Eddy-Current Inspection

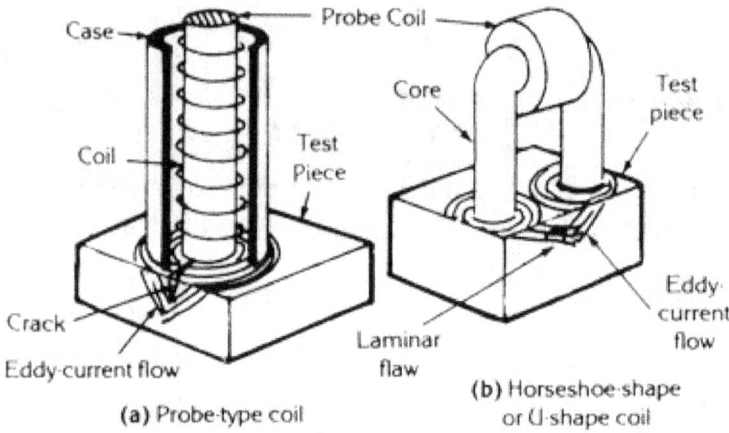

(a) Probe-type coil
(b) Horseshoe-shape or U-shape coil

> **Use:**

\# To measure thickness of non-conductive / non-magnetic coating on a conductive metal/ magnetic metal

\# To identify grain size, heat treatment condition, hardness & physical dimension

\# To measure thickness of coating on a to detect cracks, voids & inclusions

> **Advantage:**

\# Very fast

\# Can detect ultra grannular corrosion

\# Can detect pitting

\# To measure coating thickness

VIII. HARDNESS TEST

Hardness of a material determines the resistance to wear, penetration, machinability & ability to cut.

Commonly used hardness measurement methods are,

- # Brinell test: in this method, a hardened ball under load is indented on the surface of the work piece. Average diameter of the impression is measured with a low power microscope fitted with scale

- # Diameter of the ball is usually 10 mm & load of 3000 kg for steel, 1000 kg for cu and 500 kg for al.

- # Time of loading is 15 sec.

- # Thickness of the specimen should not be less than 10 times the depth of impression.

- # Vicker hardness test: in this method, a square based pyramid made of diamond under load is indented on the surface of the work piece. Average diameter of the impression is measured with a low power microscope fitted with scale

- # Load ranges from 5 kg to 120 kg.

- # Load divided by the contact area of impression gives the vickers pyramid no.(vpn).

- # As impression is small polished surface is required for measurement

Note: brinell & vicker hardnmess values are practically identical upto a hardness of 300. Brinell no. Is not reliable beyond 600.

- # Rackwell hardness test: in this method, aball of 1/16" dia. Isfirst loaded with 10 kg. Major load of 100 kg (for b-scale) / 150 kg (for c-scale) is then applied. Dial gauge in the machine records the hardness in rockwell no.

This test is useful for rapid test on finished products.

IX. **HYDROTEST**

It is the process of filling an equipment such as tank, vessel, exchanger, furnace/ boiler tubes, pipings with water / liquid at the appropriate pressure to ascertail the strength of the equipment and to check tightness against the leak.

Complete isolation of any system to be tested is necessary to prevent the testing medium to enter the connecting lines where the presence of the medium may be harmful/ undesirable/ not required.

Instrument lines, gauge glasses, control valves, pressure relief valves, expansion joints needs to be kept isolated from being pressurised.

Complete air should be expelled from the testing circuit / equipments through vents to avoid a violent failure in the system due to expansion of the compressible air rather than water / liquid.

Care should be taken to avoid pverpressuring the system.

Calibrated pressure gauges of proper range to be used. Pressurisation should be done steadily to avoid any sudden shock in the system.

Hyd. Test pressure should be 1.5 times the design pressure. Where design pressure is not known, hyd. Test pressure should be 1.5 times the pump shut off pressure/ max operating pressure of the system.

4.8 Design of Inspection Strategy –an Example

In industry, mainly in oil refinery & petrochemical sectors are looking over a problem for improper DFT(Dry Film Thickness) in large number of area (piping & tank maintained) painting system by third party. Result, consumer's risk is gradually increased.

In this research venture, developed a procedure to draw an O.C curve using Minitab software to minimize the consumer's risk on external & internal in a large area of tank & pipe line painting to resist corrosion by best expert's review using MCDM method.

The paper deals with large area coating system by measuring NDT methods (DFT) to improve the quality level as well as better inspection strategy.

Statement of the problem
Basically, in oil refinery/petrochemical based industry required large number of area(tank and pipe line) to come under painting system to resists corrosion. This type of project done by third party. To maintain the quality control by DFT checking responsibilities are vested on the department of inspection. Practically, this is not possible to 100% inspection. An inspector inspected this type of jobs by his experience and applies Standards Used for DFT Measurement. Lack of 100% inspection, improper coating found many places, Result, gradually increase the risk to failure the coating system.

Acceptance sampling is an inspection procedure used to determine whether to accept or reject a specific quantity of material. It is a part of operation management or of accounting, auditing and services quality supervision. Acceptance sampling is most likely to useful in the following situation:
1) When the time and cost of 100% inspection is extremely high.
2) When 100% inspection is not technologically feasible or would required.
3) When there are many items or spots to be inspected and the inspection error rate is sufficiently higher than 100% inspection might cause a higher percentage of defectives (here, low DFT in painting inspection) to be passed than would occur with the use of sampling plan.

Dry film thickness (DFT) is probably the single most important measurement made during inspection or quality control of protective coating application. Even the most basic protective coating specification will inevitably require the DFT to be measured. It is considered to be the most important factor determining the durability of a coating system. The thickness of each coating layer in a system and the total system DFT will

have to be measured and recorded to show that the specified system will meet the desired durability.
In industry, Standards Used for DFT Measurement: SSPC-PA 2, AS 3894.3, ISO 19840, PSPC.

Figure-1:Elcometer:DFT measurement machine

Time study is a direct and continuous observation of a task, using a timekeeping device (e.g., decimal minute stopwatch, computer-assisted electronic stopwatch, and checkingvideotape camera) to record the time taken to accomplish a task[3] and it is often used when:
- there are repetitive work cycles of short to long duration,
- wide variety of dissimilar work is performed, or process control elements constitute a part of the cycle.

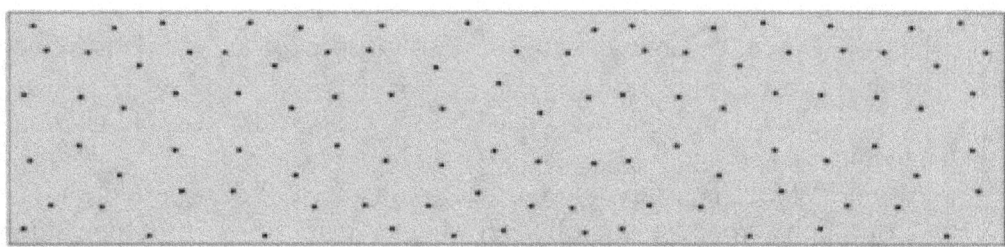

Figure -2: coating thickness measurement spot on flat area of tank plate.

Multiple criteria decision making (MCDM) is the process of selecting the best alternative from a set of feasible alternatives considering multiple conflicting criteria. In precise terms criteria are considered to be 'strictly' conflicting if the increase in satisfaction of one results in a decrease in satisfaction of the other. An MCDM process always contains at least two alternatives and two conflicting criteria (Bhattacharya et al., 2003).

MCDM are divided two broad categories: Multiple Attribute Decision Making (MADM) and Multiple Objective Decision Making (MODM).

2. Mathematical Model:

PART-1
Taken the 732 square meters area painting inspection, we, fixed the inspection spot in 100spot/sq.meter and get a lot size 73200 units

- Existing Sampling plan adapted by various experienced expert in IOCL to draw a O.C curve in this case study as given below:

SL. NO.	LOT SIZE (N)	EXPERTS	SAMPLE SIZE (n)	ACCEPTANCE NUMBER (C)
1	73,200	A	1600	50
2	73,200	B	1200	20
3	73,200	C	2000	20
4	73,200	D	1000	10

Table-1:

- For calculation of probabilities of acceptance, a Poisson's Distribution method is used as given below.
  Drawing an O.C curve on existing sampling plan:

Lot size (N) =73,200
Sample size (n) =1600,1200,2000,1000 etc.by various experts.
- Inspected the sample of lot size and if number of defectives (here, low DFT) are equal to Acceptance number(C) then accept the DFT or if it is more than C (i.e. =1, 2, 3…) then it will reject the lot.
  We have to draw an O.C curve for this assume the
  Percent of defectives in a lot as, P' (% defectives) =0.1%, 0.5%, 0.8%, 1.0%, 2.0%, 4.0%, 6.0%, 10.0%

According to Expert "A"
N=73200, n=1600,C=50
We consider, simplifying the calculation all data divided by 100 and we get
N=732,n=16,C=0

SL NO.	Percent of defectives (P')	np'	Probability of Acceptance (P_a)
1	0.1%	0.02	0.98
2	0.5%	0.08	0.92
3	0.8%	0.12	0.88
4	1.0%	0.16	0.85
5	2.0%	0.32	0.72
6	4.0%	0.64	0.52
7	6.0%	0.96	0.38
8	10.0%	1.60	0.20

Table-2: Calculation of single sampling plan for Expert-A

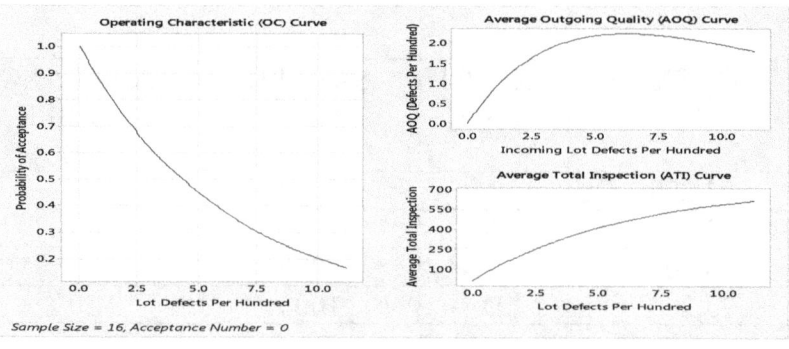

Figure-5: O.C. curve for Expert-A

If we consider lot tolerance percent defectives as 5% corresponding consumer's risk can be calculated.
From fig-1, if we consider LTPD= 5%. The Consumers risk is 44.9%

Hence consumer's risk is 44.9% with the existing sampling plan. It must be minimized.

Similarly,
According to Expert -B
N=732, n=12, C=0

SL NO	P'	np'	P_a
1	0.1%	0.01	0.99
2	0.5%	0.06	0.94
3	0.8%	0.10	0.91
4	1.0%	0.12	0.89
5	2.0%	0.24	0.77
6	4.0%	0.48	0.61
7	6.0%	0.72	0.48
8	10.0%	1.20	0.30

Table-3: Calculation of single sampling plan for Expert-B

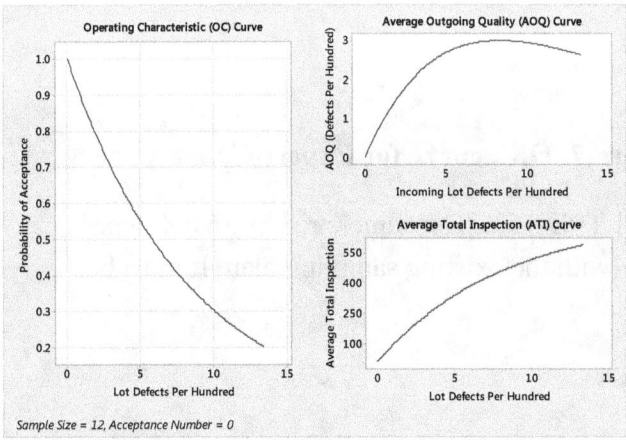

Figure-6: O.C. curve for Expert-B

From fig if we consider LTPD= 5%. Consumers risk is 54.9%

Hence consumer's risk is 54.9% with the existing sampling plan. It must be minimized.

According to Expert -C
N=732, n=20, C=0

SL NO.	P'	np'	P_a
1	0.1%	0.02	0.98
2	0.5%	0.10	0.90
3	0.8%	0.16	0.85
4	1.0%	0.20	0.82
5	2.0%	0.40	0.67
6	4.0%	0.80	0.45
7	6.0%	1.20	0.30
8	10.0%	2.00	0.13

Table-4: Calculation of single sampling plan for Expert C

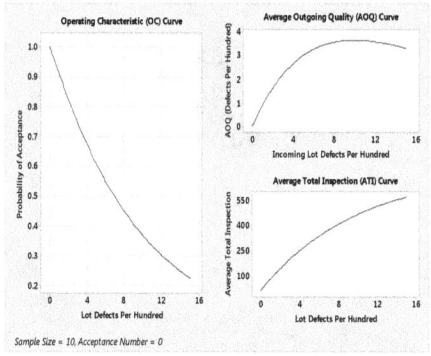

Figure-7: O.C. curve for Expert-C

From fig if we consider LTPD= 5%. Consumers risk is 36.8% Hence consumer's risk is 36.8% with the existing sampling plan. It must be minimized.

According to Expert "D"
N=732, n=10, C=0

SL NO.	P'	np'	P_a
1	0.1%	0.01	0.99

2	0.5%	0.05	0.93
3	0.8%	0.08	0.92
4	1.0%	0.10	0.90
5	2.0%	0.20	0.81
6	4.0%	0.40	0.67
7	6.0%	1.60	0.55
8	10.0%	1.00	0.368

Table-5: Calculation of single sampling plan for Expert -D

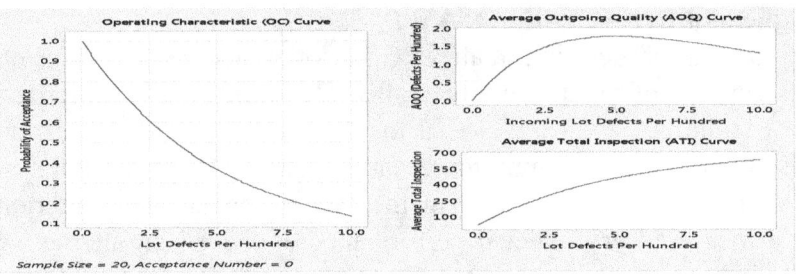

Figure-8: O.C. curve for Expert -D

From fig if we consider LTPD= 5%. Consumers risk is 60.7%
Hence consumer's risk is 60.7% with the existing sampling plan. It must be minimized.

Limitations and problems in existing sampling plan:
1. Lot tolerance percent defectives are larger.
2. As this sampling plan is based upon single sampling, there may be possibility of acceptance of defective lots.
3. As acceptance number C=0, vendors may be supply by considering this acceptance number. However

some tolerance limit should be given to the vendors for better coordination and commitment.
4. Consumer's risk in a single sampling plan is always larger and it is acceptance of such a lot, which would have been rejected. It can affect the consumer and his next production work and assembly work.

Possible Remedies for Overcoming the Limitations:

1. If increases the sample size and increases the acceptance number lot tolerance % defectives can be

Minimized. But still it is single sampling hence there may be possibility of acceptance of defective components.

2. By using single sampling inspection of various experts, comparatively analysis to determine the best minimized LTPD of this expert's decision.

3. Using the another optimization method MCDM to calculate the ranking the best expert in respect of sample size, consumer risk, and working time to minimized as well as optimized the inspection system.

PART-2
7.4.1. Simple additive weighting (SAW)

Step 1 Formation of decision matrix: Criterion outcomes of decision alternatives can be collected in a table called Decision Matrix comprised of a set of columns and rows. The matrix rows represent decision alternatives, with matrix columns representing criteria. A value found at the intersection of row and column in the matrix represents a criterion outcome - a measured or predicted performance of a decision alternative on a criterion. The decision matrix is a central structure of the MCDA/MCDM since it contains the data for comparison of decision alternatives.

$$X = \begin{array}{c} \\ A_1 \\ \vdots \\ A_i \\ \vdots \\ A_m \end{array} \begin{array}{c} C_1 \quad\quad C_J \quad\quad C_n \\ \begin{bmatrix} x_{11} & \cdots & x_{1j} & \cdots & x_{1n} \\ \vdots & \cdots & \vdots & \cdots & \vdots \\ x_{i1} & \cdots & x_{ij} & \cdots & x_{in} \\ \vdots & \cdots & \vdots & \cdots & \vdots \\ x_{m1} & \cdots & x_{mj} & \cdots & x_{mn} \end{bmatrix} \end{array}$$

.........(1)

x_{ij} is the performance rating of alternative i with respect to criterion j,
A_j is ith alternative, C_j is the jth criterion

Step 2 Formation of Weight Matrix:

Different importance weights to various criteria may be awarded by the decision makers. These importance weights forms the weight as follows.

$W = [W_1 \cdots W_j \cdots W_n]$ ...... (2)

Step 3 Normalization of performance rating

Units and dimensions of performance ratings of columns under criteria differ. For the purpose of comparison, these performance ratings are converted into dimensionless units by normalization using following equations:

$$\bar{x}_{ij} = \frac{x_{ij}}{\max_i(x_{ij})} \quad \text{for benefit criteria} \quad \ldots\ldots(3)$$

$$\bar{x}_{ij} = \frac{\min_i(x_{ij})}{x_{ij}} \quad \text{for non-benefit criteria} \ldots(4)$$

Normalized decision matrix

Step 4: Composite score:
Computation of composite score (CSi) for alternatives.

$$CS_i = \sum_{j=1}^{n} (\bar{w}_j * \bar{x}_{ij})$$

Step 5 Ranking and selection of best alternative:

Ranking of products in descending order of composite scores (CSi)

$$\bar{X} = \begin{matrix} A_1 \\ \vdots \\ A_2 \\ \vdots \\ A_m \end{matrix} \begin{bmatrix} \bar{x}_{11} \cdots & \ldots \bar{x}_{1j} \ldots & \bar{x}_{1n} \\ \vdots & \vdots & \vdots \\ \bar{x}_{i1} \cdots & \ldots \bar{x}_{ij} \ldots & \bar{x}_{in} \\ \vdots & \vdots & \vdots \\ \bar{x}_{m1} & \bar{x}_{mj} & \bar{x}_{mn} \end{bmatrix}_{m \times n} \quad \ldots\ldots\ldots(5)$$

> . **Formation of Matrix comparing with other Inspection Factors:**

SL. NO.	SAMPLE SIZE(n)	CONSUMER RISK (L.T.P.D) [5%]	TIME (TIME STUDY METHOD)

EXPART-A	16	44.9%	7min.
EXPART-B	12	54.9%	5.4min.
EXPART-C	20	36.8%	9min.
EXPERT-D	10	60.7%	4.5min.

Table-6

> Computational Result by Mat lab software:

The weighted values are:

Criteria-1	Criteria-2	Criteria-3
0.3381	0.3222	0.3397

Table-7: weighted values by Entropy method

✓ Result of the SAW method:

The values of (s) are:

A	B	C	D
0.7529	0.7019	0.8301	0.7041

Table-8: Final result in ranking

Arranging the final value in descending order: E3>E1>E4>E2

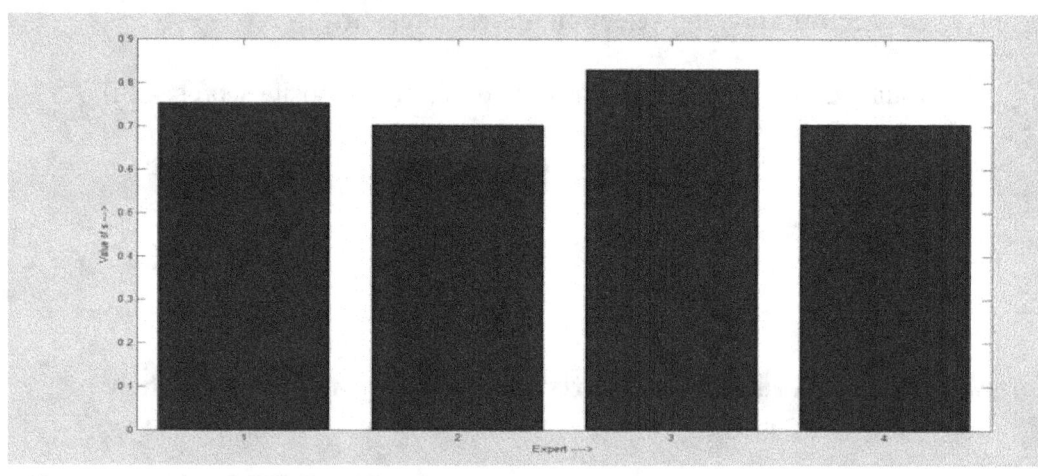

Figure-9: Ranking best Expert by MCDM.

Conclusions:
The design of acceptance sampling process of this particular job includes decisions about sampling versus complete inspection. In this design strategy has shown prominently the quality risk through the graphs individually for each inspection experts, using Minitab software, that can easily and more quickly calculate and draw operating characteristic curve.
The second level optimization tool,MCDM is used for selection the best expert comparing with other factors using Mat lab software that can easily shown the graph for best decision through expert's ranking.
This model helps the large number of painting inspection become automated as well as less time consuming.

4.9 IE approach in NDT environment-An Example

In industrial storage tank, specifies minimum acceptable thicknesses for the bottom & annular plate [Figure: 1] by API 653. It indicates that the remaining thicknesses may be quantified using either probabilistic or deterministic method. The bottom plate thicknesses are then compared to the required thicknesses to determine their acceptability.

The deterministic method uses more extensive inspection data to quantify the remaining thickness of the bottom. The required data include the following:

- Original plate thickness.
- Average and maximum depth of internal and underside pitting.
- Maximum internal and underside pitting rates and maximum general corrosion rate.
- Average depth of general corrosion.
- The time until the next inspection.

But in practical approaches of oil & petrochemical industry, there are various manual inspection problems occur for the hazardous condition like Tank, Furness, boiler etc.

Often it is observed that the report was not correct. For better decisions, introducing the inspection accuracy on alternative/cross check policy should be implemented by execution to analysis the inspection accuracy.

The model uses the fraction defect rate in the inspection batch, the cost of inspection per item which is inspected, and the cost of damage that one defective plate would cause if it were not inspected properly. The total cost per plate 100% inspection can be formulated.

Figure: 1

1.1 Inspection accuracy: Refers to the capability of inspection process to avoid these types of errors; Measures of inspection accuracy are suggested by DRURY for the case in which parts are classified by an inspector. Inspected items of good quality are incorrectly classified as not conforming to accepted specification, and nonconforming items are mistakenly classified as conforming. These two kinds of errors are called "False alarm" and "Miss" respectively.

Two types of errors in cross check inspection:

	Conforming Item	Nonconforming Item
Accepted Item	Right decision	Miss
Rejected Item	False alarm	Right decision

1.2 Inspection or No inspection: A model for deciding to inspect at a certain point in the production sequence is proposed in Juran and

Gryna. The model uses the fraction defect rate in the inspection batch, the inspection cost per unit inspected, and the cost of damage that one defective unit would cause if it were not inspected.

2. STATEMENT OF THE PROBLEM

During tank maintenances, the bottom plate metal loss checking is an important part of tank inspection. Sometimes error occurs in the inspection procedure such as those plates of good quality are incorrectly classified and vice-versa. In manual inspection, these errors result from factors such as:

i. Inherent variations in the inspection procedure
ii. Complexity, hazard and other various difficulty of the inspection task
iii. Inaccuracy in measuring instrument
iv. Mental fatigue etc.

For this irresolute inspection errors, It's have possible chance to leak [Figure: 02 & 03] of the tank before the next tank Maintenance & Inspection (M & I), results-

- Production loss
- Drain of money
- Idle times are increase etc.

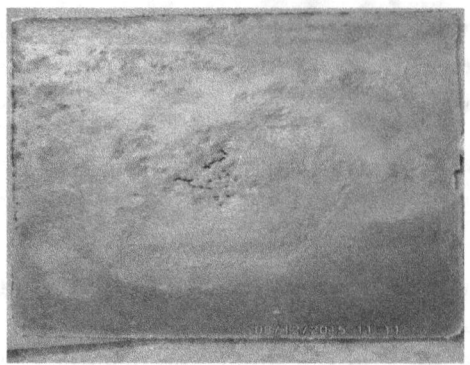

Figure: 2 Failure sample bottom plate

Figure: 3 Failure sample annular plate

On this critical inspection cases the inspection accuracy is very important. But these types of situation the exact accuracy as well as defect rate remain undetermined. Whenever, the leak is found for failure of inspection then it is impossible to assess the past inspection accuracy on the basis of inspected data.Modeling a system to make automated inspection accuracy for minimizing the decision criticality. For minimizing this type of errors, implemented the alternative/cross check inspection are frequently used [Figure: 4].

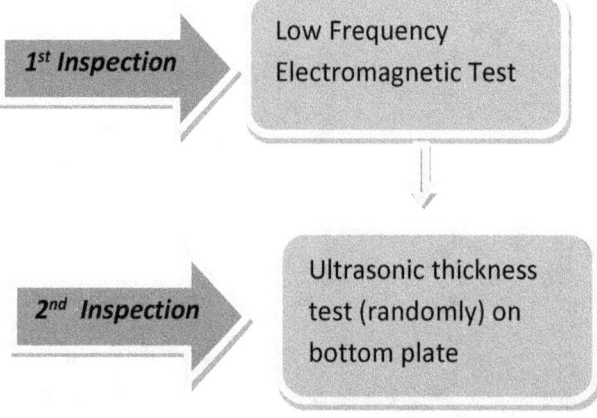

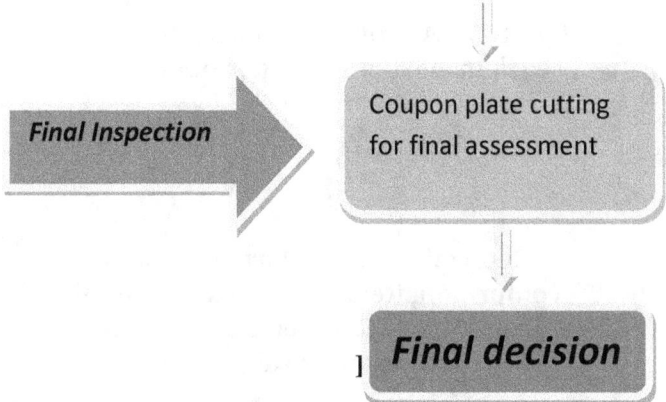

3. RESEARCH OBJECTIVES

In practical approaches, during LFET on bottom floor of a tank inspection, frequently errors are come out for hazarded condition. To avoid this kind of situation a different approach has been implemented i.e. coupon plate cutting for sample inspection, Thickness checking on spot sampling etc. As a result, processing time of inspection is maximized, risk factor is unpredictable, No future inspection can be done because of increasing the inspection cost etc.

In this research, addresses to analysis the inspection accuracy by

I. Automated computational methodology to determine the accuracy in cross check approaches.
II. Finding the errors for future inspection
III. Simplifying the decision about the plates will be replaced or not.

4. MODELING INSPECTION SYSTEM FLOW DIAGRAM

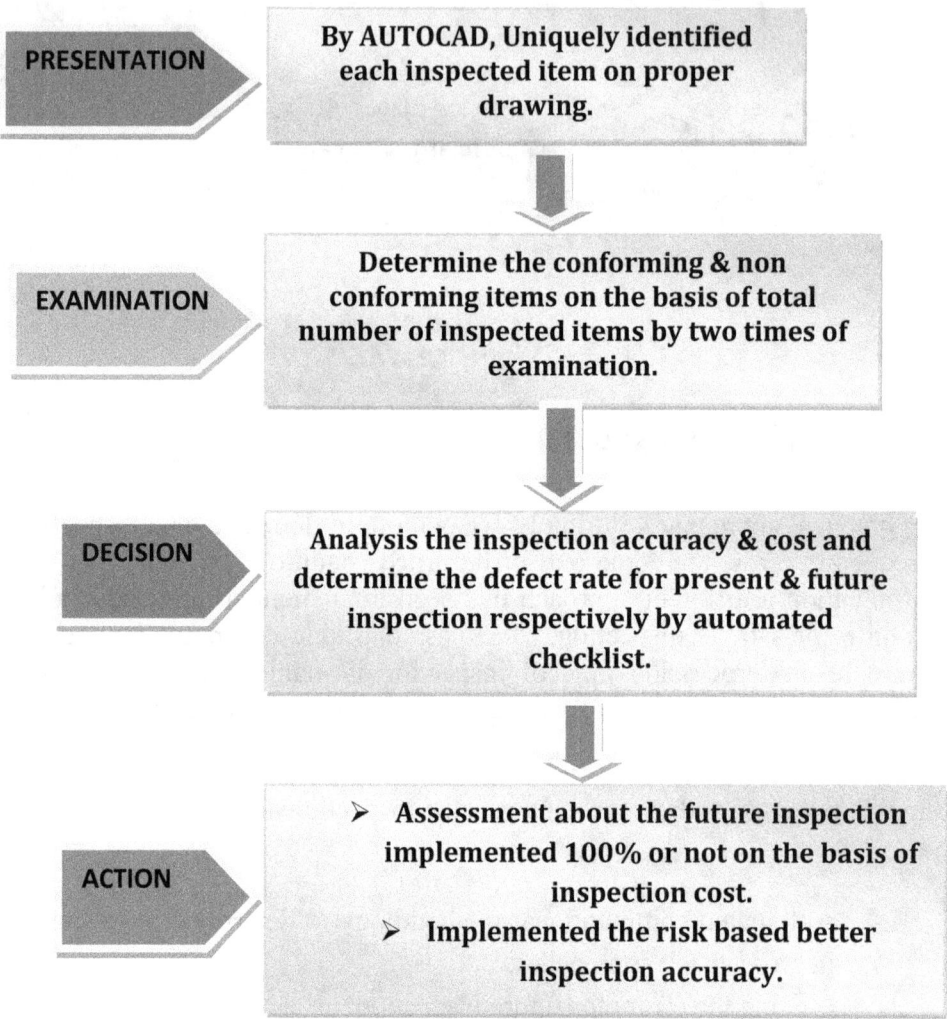

5. EXPERIMENT

Basically in oil & petrochemical industry looking over an inspection problem in hazarded area like Tank, Furnace, boiler etc. During manual inspection, various types of errors are occurring in inspection time and the inspection report become incorrect by human error. To make the inspection accurate the cross check policy should be implemented by execution.

During storage tank maintenance, the bottom plate thickness test is a part of critical inspection, because various types of hazard are occurred that time. Analysis of the inspection accuracy on bottom plate [Figure: 05] thickness during tank M&I under cross check methodology:

5.1 Design & Identify the each inspected plates by numbering through the CAD on complete bottom area.

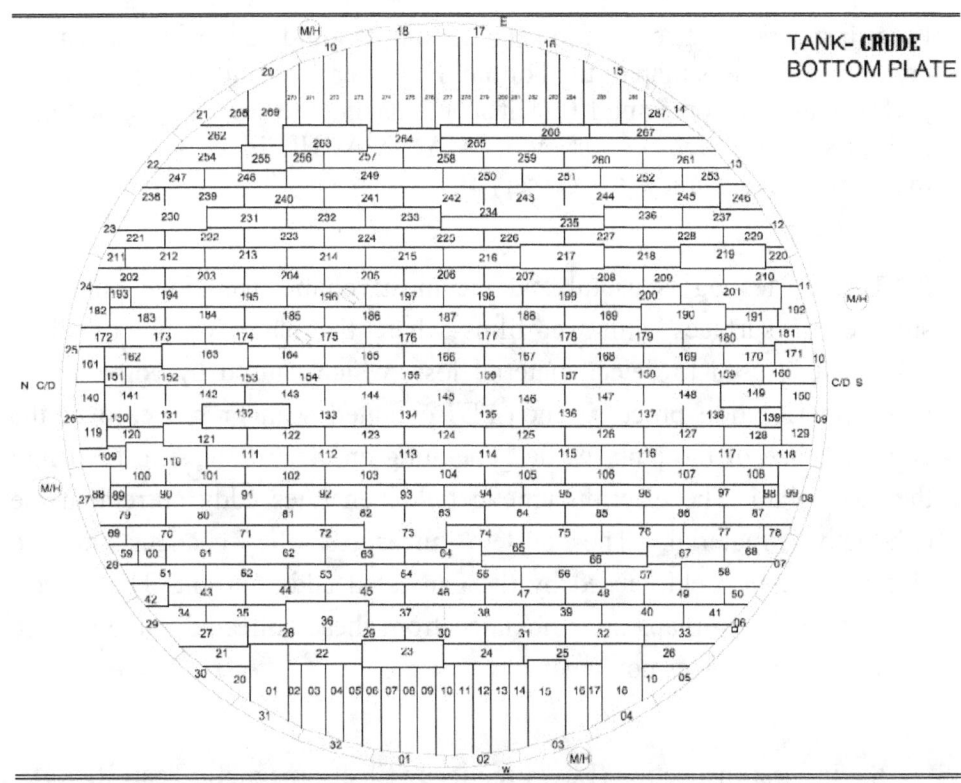

Figure: 05

Using the following drawing:

- Total bottom plates of tank = 287
- Total annular plates of tank = 32

So, Total number of inspected plates of tank = 319

- ✓ Each plate inspection cost (C_S) = INR 950 [approx]
- ✓ Each plate damage cost (C_d) = INR 38,000 [approx]

5.2 Inspection and Data Analysis

From inspection data evaluation standpoint, it would be ideal to have a complete thickness map of the bottom. However, it would be expensive to perform the ultrasonic and LFET inspections that are necessary to do that, and for such an extensive survey of "MAN-MACHINE" error for analysis the inspection accuracy is necessary.

5.2.1 LFET testing is useful and the most reliable technique on rough surfaces or surfaces with wet films where the plates or the pipelines coated. It is used to detect material loss, which caused by corrosion or other deterioration process. The LFET operated scanner moves over the entire surface of the tank, while generating an electromagnetic field into the steel plate. The electromagnetic field generates eddy current in the conductive material. The system measures the changes in the electromagnetic field caused by the generated eddy current. The defects and the corrosion maps are calculated from these collected values. Since the scanner head does not have

- ❖ In First Inspection (Low Frequency Electromagnetic Test) Report,
  The number of defective plates are detected = 56
  The percentages of good or defective items are shown in following chart [Figure: 06]

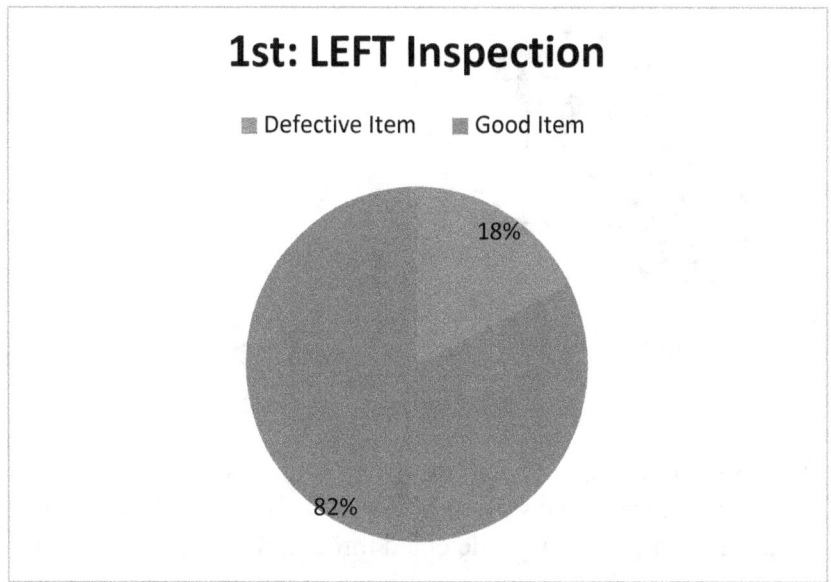

Figure: 06

In hazarded or critical based inspection condition, the second inspection is implemented to minimize the risk.

5.2.2 Ultrasonic nondestructive testing (NDT) – a method of characterizing material thickness, integrity, or other physical properties by means of high frequency sound waves is a widely used technique for product testing [Figure: 8] and quality control. In thickness gauging applications, ultrasonic techniques permit quick and reliable measurement of thickness without requiring access to both sides of a part. Calibrated accuracies as high as ±2 micrometers or ±0.0001 inch are achievable in some applications. Most engineering materials can be measured ultrasonically under proper maps to evaluate the thickness data [Figure: 7] .

Figure: 7

❖ In second Inspection Report, it was found that 16 of these reported defects were in facts good pieces. Whereas a total of 21 defective plates in a tank were undetected through the inspection [Figure: 9]

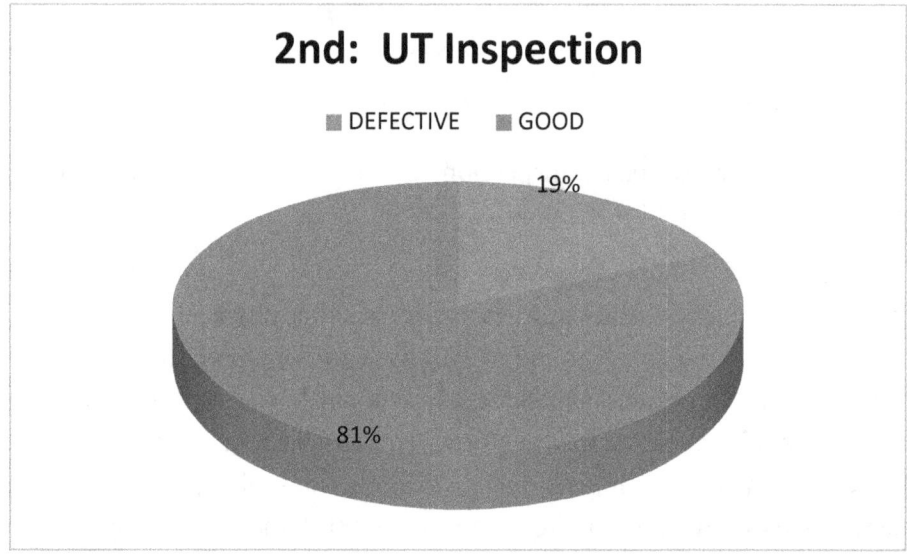

Figure: 9

➢ Comparative analysis of inspection data in two times different inspection procedure [Figure: 10]

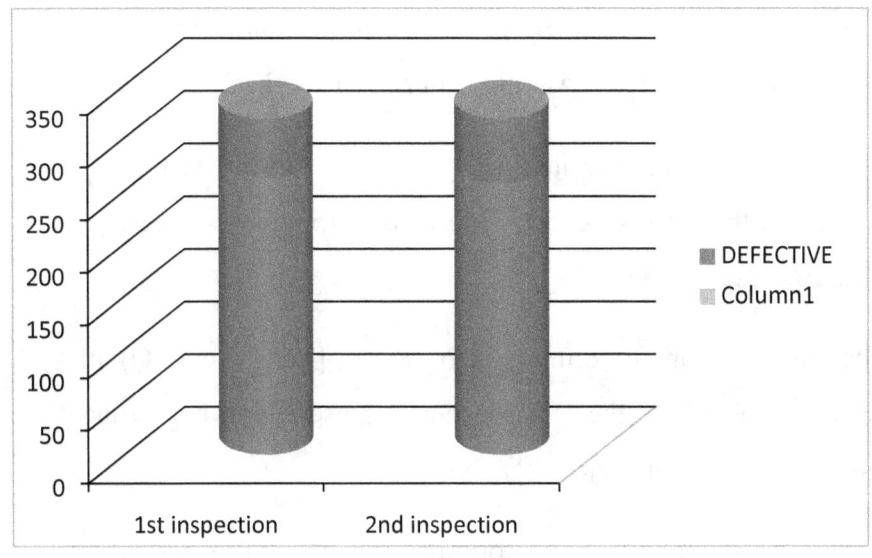

Figure: 10

So, the total "FALSE ALARME" = 16

The total "MISSES" = 21

5.3 Constructing computing logic on the inspection data by Microsoft Excel as following

Microsoft Excel has the basic features of all spreadsheets, using a grid of cells arranged in numbered rows and letter-named columns to organize data manipulations like arithmetic operations. It has a battery of supply functions to answer statistical, engineering and financial needs.

Developed a check sheet by Microsoft Excel [Figure: 11]:

- Actual Acceptance (A) = Good (G) in the 1st inspection + False Alarm (F) during cross check – Miss (M) during cross check.

- Actual Reject (R) = Bad (B) in the 1st inspection + Miss (M) during cross check – False Alarm (F) during cross check

- Probability of conforming item (P_1) = [Total Item (Q) in batch – {Bad (B) in the 1st inspection + Miss (M) during cross check}] / Actual Acceptance (A)

- Probability of Non-conforming item (P_2) = [Total Item (Q) in batch – {Good (G) in the 1st inspection + False Alarm (F) during cross check}] / Actual Reject (R)

- Accuracy = {Probability of conforming item (P_1) + Probability of Non-conforming item (P_2)} / 2

- Defect Rate (q) = (1 - over all inspection accuracy)

- Batch cost for 100% inspection (C_b) = $Q \cdot C_S$ (Q= total parts in a batch) &

  Cs = inspection cost

- Batch cost for NO inspection (C_n) = $Q \cdot q \cdot C_d$ (C_d = inspection damage cost)

- The critical defect value (Q_C) = C_S / C_d [critical value represents the breakeven point between inspection or no inspection]

Total Parts in a Batch	1st inspection			Cross Check		Good Decision		Actual Probability of conforming item	Probability of Non-conforming item	Accuracy	Defect Rate	Inspection Cost	Damage Cost	Batch cost for 100% inspection	Batch cost for NO inspection	the critical defect value
	Good	Bad	False Alarm	Miss	Actual Acceptance	Actual Reject										
Q	G	B	F	M	A	R	P1	P2	A	q	Cs	Cd	Cb	Cb	Qc
100	88	12	4	6	86	14	0.953488372	0.571428571	0.76245847	0.23754153	10	15	1000	356.3122924	0.66666667
20	10	10	6	4	12	8	0.5	0.5	0.5	0.5	6	9	120	90	0.66666667
319	263	56	16	21	258	61	0.937984496	0.655737705	0.7968611	0.2031389	950	38000	303050	2462449.739	0.025

Figure: 11

5.4 RESULT

- The proportion of good parts reported as conforming is $(P_1) = 0.938$
- The proportion of defective parts reported as nonconforming is $(P_2) = 0.656$
- The overall inspection accuracy $(A) = 0.797$ [Figure: 12]
- Defect Rate $(q) = 0.203$

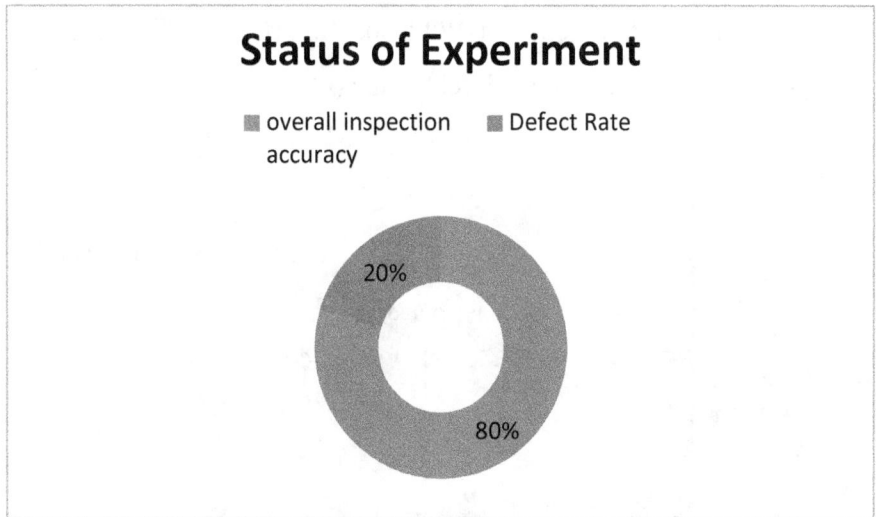

Figure: 12

- In quality assurance approaches, if we consider the overall inspection accuracy is an average accuracy then it show the inspection status as a graph [Figure:03] in Poisson distribution by Minitab software as

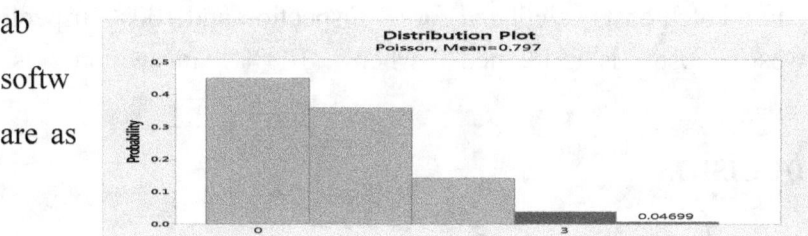

following:

Figure: 03

I. Batch cost for 100% inspection (C_b) = INR 3,03,050
II. Batch cost for NO inspection (C_n) = INR 24,60,766
III. The critical defect value (Q_C) = 0.025

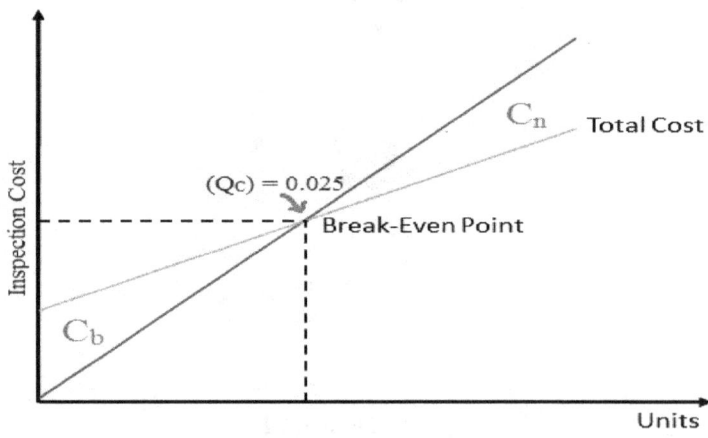

Figure: 04 Break even Analysis

Since the anticipated defect rate in the inspected batch is = 0.203, the decision should be inspect. Observed that this decision is consistent with the two batch costs calculated for no inspection and 100% inspection. The lowest cost is attained when 100% inspection is used.

5.5 DECISION

Based on past history with the inspected items, the batch fraction defect rate q is less than this critical level then no inspection is indicated. On the other hand, if it is expected that the fraction defect rate will be greater than Q_C, then further inspection is necessary.

If, Q_C < q	Inspection is indicated
If, Q_C > q	NO inspection is indicated

In this experiment,

The critical defect value (Q_C) = 0.025 is less than the Defect Rate (q) = 0.203

$$Q_C < q$$

❖ Future or further inspection is indicated

5.6 ADVANTAGES OF THIS METHODOLOGY

- Determine the inspection accuracy easily in excel sheet.
- Assessment of the inspector's responsibility on the particular job
- Analysis the risk based inspection report
- Inspections are less time consuming
- Determine the decision about future inspection under cost based inspection.
- Minimize the human error

CONCLUSION

The inspection methodology is used to determine the error rate for quality control during critical inspection environment. It also helps to evaluate the risk for future inspection by automated data record. The risk based inspection policy is merged on cost based inspection methodology to analysis the next inspection criticality. The complete execution process is made on computationally as well as less time consuming. This model useful for various inspection procedures as well as prediction of human error.

References:

[1] Juran. (2002) Quality control handbook. McGraw-Hill Books

[2] Mahajan.M., "Statistical Quality Control" Edition-1998,Dhanpat Rai & Sons, India

[3] Telsang.M.,"Industrial Engineering and Production Management" Second Edition, S.Chand & Company Ltd.,2012,ISBN-81-219-1773-5.

[4] DRURY,C.G., "Inspection Performance, " Handbook of Industrial Engineering, Second Edition,G.Salvendy,Editor,John Wiley & Sons,Inc.,New York,1992,pp 2282-2314.

[5] Juran, J.M., and Gryna, F.M., Quality Planning and Analysis, Third Edition, McGraw-Hill, Inc., New York, 1993.

[6] Tannock, J.D.T., Automating Quality Systems, Chapman & Hall, London,1992.

[7] Groover, M.P, "Automation, Production Systems, and Computer-Integrated Manufacturing" Second Edition, Pearson Education, Inc., 2007,ISBN-81-317-0227-8.

[8] Jana P,Dutta M,A.K.Kar: "Design an Inspection Strategy on DFT Checking to Determine the Quality Risk through the O.C. Curve on Single Sampling Plan using the Best Expert Decision under MCDM Environment" Imperial Journal of Interdisciplinary Research (IJIR), Vol-2, Issue-8, 2016 (ISSN-2454-1362),831-837.

[9] API Recommended Practice 652, Lining of Aboveground Storage Tank Bottoms

(Washington, DC: American Petroleum Institute, 1991).

[10] API Standard 653, Tank Inspection, Repair, Alteration, and

Reconstruction (Washington, DC: American Petroleum Institute, 1995).

[11] Vasadi Vishwesh et al. "COST EFFECTIVE AND EFFICIENT INDUSTRIAL TANK CLEANING PROCESS" International Journal of Research in Engineering and Technology (IJRET), Vol:03, Issue: 03, Mar-2014(eISSN: 2319-1163,pISSN: 2321-7308),450-454.

[12] Prithwiraj Jana et al. "Analysis of automated risk based inspection cost & accuracy on industrial tank maintenance & other critical NDT aspects: An executive approach" WSN 55 (2016) 274-288

[13] *Media reports, Press releases, EEPC India, Press Information Bureau (PIB), Department of Industrial Policy and Promotion (DIPP)*

Authors Biography

P.Jana born in India 1990. Obtained his Bachelor's degree in Production Engineering, from Haldia Institute of Technology, during 2008-2012. & Master's degree from West Bengal University of Technology in the Industrial Engineering & Management during 2012-2014.He is having about 02 years industrial experience in I.O.C.L and having 05 International journals / Conference papers. He also obtained his professional qualification on NDT ASNT (The American Society for Nondestructive Testing) Level-II (UT, DPT, MPT, RT). He is also an author of various Engineering books and magazines.

www.ingramcontent.com/pod-product-compliance
Lightning Source LLC
Chambersburg PA
CBHW082248220526
45469CB00009B/2919